新 发 展

中国传统文化

经典教程

New Development

Course of Classics for Traditional Chinese Culture

特邀主审　黄大乾

总 主 编　何高大

主　　编　唐红梅

副 主 编　肖光俊　刘　岚

　　　　　傅　佳　张志伦

北京理工大学出版社
BEIJING INSTITUTE OF TECHNOLOGY PRESS

图书在版编目（CIP）数据

新发展中国传统文化经典教程 / 唐红梅主编 . -- 北
京：北京理工大学出版社，2021.8
ISBN 978-7-5763-0163-2

I.①新… Ⅱ.①唐… Ⅲ.①中华文化 – 英语 – 高等
学校 – 教材 Ⅳ.① K203

中国版本图书馆CIP数据核字(2021)第165609号

出版发行 / 北京理工大学出版社有限责任公司
社　　址 / 北京市海淀区中关村南大街 5 号
邮　　编 / 100081
电　　话 /（010）68914775（总编室）
　　　　　（010）82562903（教材售后服务热线）
　　　　　（010）68944723（其他图书服务热线）
网　　址 / http://www.bitpress.com.cn
经　　销 / 全国各地新华书店
印　　刷 / 沂南县汶凤印刷有限公司
开　　本 / 889 毫米 × 1194 毫米　1/16
印　　张 / 13.5　　　　　　　　　　　　责任编辑 / 王晓莉
字　　数 / 308 千字　　　　　　　　　　文案编辑 / 王晓莉
版　　次 / 2021 年 8 月第 1 版　2021 年 8 月第 1 次印刷　责任校对 / 刘亚男
定　　价 / 84.00 元　　　　　　　　　　责任印制 / 施胜娟

PREFACE

《新发展中国传统文化经典教程》（*New Development: Course of Classics for Traditional Chinese Culture*）是当代每位大学生的必修课程。了解并熟知中国优秀传统文化，是实现立德树人的根本，有助于培养具有家国情怀、积极进取、勇于创新的时代英才；有助于增强大学生文化自觉和文化自信，铸造国家和民族发展之魂。本教程以中国传统文化为主线，遵循由浅入深的认知规律，以任务、问题、成果目标为教学导向，实现以文化人、以文养人、以德树人的教学成果，激发学生的参与意识。同时，本教程集思政性、人文性、思辨性、工具性、实践性和实用性为一体，从多维角度为不同层次的学生提供文化给养，帮助学生了解基本的文化内涵和社会礼仪，增强文化自信，提升爱国情怀，鼓励学生在全面深刻理解"四个自信""四个意识""两个维护"的基础上，提高用英语讲好中国故事的能力。

一、编写依据

本教程以党的十七届六中全会通过的《关于深化文化体制改革推动社会主义文化大发展大繁荣若干重大问题的决定》、习近平总书记"把立德树人作为教育的根本任务"的重要论述、《高等学校课程思政建设指导纲要》和《国家职业教育改革实施方案》为指导，运用系统论、整体论和金课理念，实现学习、活动、知识、经验、理论与职业岗位能力、任务相统一，凸显金课的高阶性、创新性和挑战性，使国学文化、语言学习和职业教育实现知行合一的立体学习目标，增强学生对中国传统文化的自信，同时提升学生英语语言运用能力和职业岗位能力。

《中共中央、国务院关于深化教育改革全面推进素质教育的决定》指出要积极发展高等职业教育，大力推进职业院校教育人才培养模式的改革。本教程以中国优秀传统文化为依托，以中国哲学思想为主线，通过直观、互动的教学活动将文化融入语言学习，又将语言学习融入文化传播，使语言服务于文化、思想和实践，实现"学中做、做中学"的教育理念，拓宽学生的文化知识，提升学生的语言能力和社交能力。

二、教材特色

本教程以思政理论为纲，以经典文化为本，以古代哲学思想为主线（主要包括儒家思想、道家思想和阴阳理论），以传播中国优秀文化为目标导向，以发展思维能力为核心素养目标，夯实软实力，扩延格局，增强学生的文化自觉、文化自尊和文化认同，提升学生的自主学习能力、语言运用能力、社交能力、社交礼仪和岗位能力，从而增强学生在跨文化交际中的文化自信，培养他们辩证地看待事物发展的能力，为他们职业生涯及自身的可持续性发展打下坚实的基础。

1. 以文化人，凸显课程思政性和人文性

本教程以任务、问题和成果目标为导向，以中国传统优秀文化为依托，以立德树人为根本任务，培养学生弘扬家国情怀，提升学生思政理论休养和文化素养，使学生做到知、情、意、行的统一。通过学习中国传统文化的渊源和发展历史，学生能够了解传统文化的内涵，掌握事物发展的规律，通晓天道，丰富学识，塑造品格，能够成为德智体美劳全面发展的社会主义建设者和接班人，最终实现立志报国的伟大抱负。编者根据每单元的主题发掘设计出培养学生政治、思想、文化素养的教学活动，使其通过学习建立高尚的情操、优秀的品德、强烈的社会责任感以及深厚的文化基础和素养。

2. 以交际为导向，提升学生传播中国传统文化的能力

本教程以"文化意识—阅读先行—读写跟上—思辨随行—礼仪到位"的顺序展开，目的是激活学生原有的文化知识，扩大其语言输入，并培养其言说能力和思辨能力，引导学生能自豪地运用恰当的英语讲好中国传统文化故事，提升文化自信。

3. 以言促思，启迪学生思辨和创新

本教程坚持弘扬中国传统优秀文化，培养学生明辨是非、勤于思考、善于发现的能力，并培养学生大担当、大格局、大气魄、大胸襟的人格。每个单元根据主题，以问题和成果目标为导向，使教学任务具体化、可视化、操作化和目标化，促进全员参与课程学习，以培养知晓中国文化精髓、具备高阶思维能力的新一代大学生。

三、单元模块设计

本教程共有10个单元，每个单元主要由Brainstorming, Story Reading, Let's Practice, Extended Reading, Insight into Proverbs和Self-assessment Checklist 6个模块组成，涉及10个文化经典内容，与当代大学生的学习、生活和未来工作紧密相关。选文难度适中，有助于学生夯实语言基础，提高语言综合运用能力、思辨能力和传播文化的能力。

Brainstorming以短视频、图片开篇，具体化、可视化地引导学生进入单元主题和情境，启发学生的思维，激发学生的学习兴趣。

Story Reading以故事、趣闻或典故等为内容，融入听、说、读、写、译多种学习方式，使学生从多方位展开学习。模版设计由浅入深、循序渐进，特别是词汇表的设计有助于学生积累词汇、掌握与该主题相关的术语表达。此外本教程的练习内容丰富多样，有助于解决学生在传播文化中会遇到的实际问题。

Let's Practice以单元主题内容展开和提升交际能力为目的，结合音视频等多媒体教学手段，使学生了解所学内容在社交中的相应礼仪，进而提升学生的语言综合运用能力、跨文化交际能力和思辨能力。

Extended Reading是对单元主题的延伸和拓展，加深学生对单元主题相关文化的了解，选材注重趣味性和实用性。

Insight into Proverbs以中国传统文化中的五常五德"仁义礼智信，忠孝节勇和"的谚语为主要内容，以中英文对照的方式呈现两种语言的差异，体现中国传统文化的价值观和世界观，凸显文化中蕴含的思政性和思辨性。

Self-assessment Checklist围绕单元主题进行综合评价，让学生自我检测掌握相关文化知识、运用语言传播文化、提升交际能力的情况。

本书由何高大教授担任总主编，负责教程的总体策划、体例设计、部分内容编写、统稿、审校、智慧云、APP设计等组织管理和编写工作，唐红梅副教授担任主编，肖光俊老师、刘岚副教授、傅佳、张志伦老师担任副主编。广州大洋教育信息技术有限公司总裁周春翔先生、广东头狼教育科技有限公司朱加宝董事长、广州熙然化妆品有限公司朱昭辉董事长、唯品会（中国）有限公司高级商务经理田静女士等对本书的编写提供了许多来自岗位一线的案例和语料，并提供了诸多建设性建议，在此表示对他们真诚的感谢！

在编写过程中，我们还参考了诸多文献和网址，限于篇幅，恕未能一一列出，在此表示真诚的谢意！

由于我们水平有限，本教程难免存在纰漏，敬请读者和相关院校在使用过程中给予关注和批评指正，以便及时修订和加以完善，在此表示真诚的感谢！

编　者

2021年8月

CONTENTS

Unit 1

Ancient Chinese Education and Philosophy

China has long been known as a country with ancient civilization, and a land of ceremony and propriety. The ancient Chinese philosophy formed in the Spring and Autumn Period and the Warring States Period, a time of great social change, with different schools and ideas of pre-Qin philosophers. The philosophers wrote books and discussed the current politics, thereby criticizing and influencing each other. The main schools in the debate were from different fields, mainly including Confucianism, Taoism, Mohism, Legalism, ect. This promoted the social change and cultural development at that time. The ancient Chinese philosophy has its own unique and traditional concepts. All such concepts incorporate the wisdom of the Chinese ideologists. Then how much do you know about ancient Chinese philosophy? What's the doctrine of each school? How did these doctrines influence ancient Chinese people's life? What do they affect Chinese people's life in modern times? This unit will show you the wisdom of the ancient Chinese philosophy.

Overview

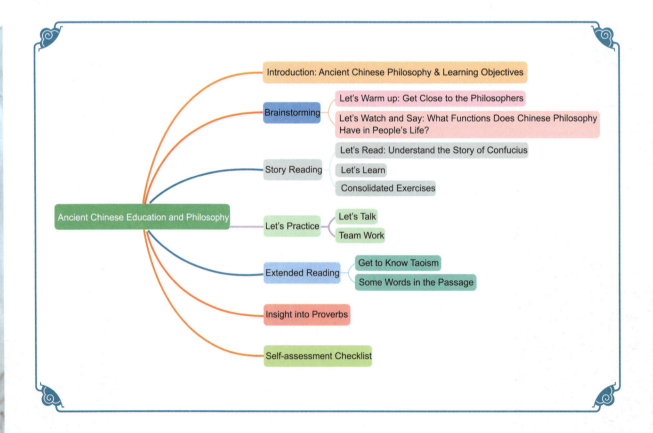

Ancient Chinese Education and Philosophy

- Introduction: Ancient Chinese Philosophy & Learning Objectives
- Brainstorming
  - Let's Warm up: Get Close to the Philosophers
  - Let's Watch and Say: What Functions Does Chinese Philosophy Have in People's Life?
- Story Reading
  - Let's Read: Understand the Story of Confucius
  - Let's Learn
  - Consolidated Exercises
- Let's Practice
  - Let's Talk
  - Team Work
- Extended Reading
  - Get to Know Taoism
  - Some Words in the Passage
- Insight into Proverbs
- Self-assessment Checklist

Learning Objectives

After learning this unit, students are able to reach the goals below.

1	专业能力目标	① 了解中国教育、中国哲学思想的起源；了解中国教育的理念及其代表人物对当今世界的影响
		② 学会与教育话题、中国古代哲学思想相关的词汇和句型
		③ 提高阅读速度；掌握文章的主旨和大意；了解孔子的生平、孔子的教育理念和政治主张及其对中国社会乃至世界文化的影响；用较流利的语言传播中国儒家思想
		④ 了解老子的思想对中国文化及世界哲学的影响；辨别儒家思想和道家思想的共性和差异
2	方法能力目标	① 学会通过主题句猜测段落大意
		② 学会使用同位语结构；懂得人物、事物的刻画技巧；提高概念或词义的理解能力和推测能力
		③ 理解动名词在句子中的含义并掌握其用法；增加句式表达的多样性
		④ 学会通过阅读上下文猜测单词的含义；懂得同义词的用法；扩大词汇量
3	社会能力目标	① 通过主题阅读，学会思考中国古代哲学思想对当代社会的影响；能运用中庸思想达到人与社会的和谐相处； 能运用哲学思想解决学习、生活和工作中的问题，从而提高社会适应能力
		② 在信息技术时代，提高获取中西方哲学思想的能力；学会分析不同的哲学思想
		③ 掌握基本的英语阅读技能，提高学习效率和社会工作效率
4	情感与思政目标	① 通过本单元的学习，提升对中国教育和中国哲学思想的认识；获得正确看待事物的判断力和决策力，成为有思想的当代大学生
		② 通过主题阅读，把握中国哲学的基本脉络；成为有理、有义、有情、有爱、维护社会和谐的合格公民
		③ 通过主题阅读及拓展阅读，学会判断中西哲学"和而不同"的基本观点；学会运用哲学思想解决问题，达到终身学习

Task 1　Let's Warm Up

　　Look at the following pictures to identify who they are and write their names on the lines. Match them with their ideas or main doctrines.

_____

A. His main doctrines are "the goodness of human nature" "Serious efforts must be made to recover men's original nature." and "The end of learning is none other than to 'seek for the lost mind'."

_____

B. He advocated the natural law of "The Tao gives birth to One. One gives birth to Two. Two gives birth to Three. Three gives birth to all things." Things in opposition will transform into each other, that is, the transformation of Yin and Yang. All things follow this way, so he advocated non-action.

_____

C. He advocated benevolence and courtesy among people, which helped to shape the value of sacrificing one's personal interests for the good of others and the collective among Chinese people.

_____

D. Putting forward the theory of "escaping from the world", he advocated the obedience to the "The Dao of Nature" and the communication between heaven and earth. He said that the real Tao and perfect love and benevolence are no-self and being flexibility.

Task 2   Let's Watch and Say

1 Before watching the video, think about the following questions.

(1) What is philosophy in your mind?

(2) How much do you know about ancient Chinese philosophy?

(3) Do you think Chinese philosophy can change your life?

2 After watching the video, arrange your mind and make a conclusion of the above questions in pairs.

Section II
Story Reading

Task 1　Let's Read

Confucius

Born in 551 BC, Confucius, the greatest teacher in the history of China, was named Kong Qiu and would someday be called Kong Fuzi. Confucius' father, Kong Shulianghe, was a descendant of the royal family of Shang and had at one time been governor of the town of Zou. When Kong Qiu was three years old, his father died, leaving the child and his young mother in poverty. As Kong Qiu grew older, their financial situation required him to take on menial jobs such as caring for livestock. He became known as a polite, smart and hard-working young man with an insatiable desire to learn. This thirst for knowledge would push him toward a life of learning and teaching others.

1. His Educational Career

At the age of fifteen, Confucius began to devote himself to serious studies and made up his mind to become a scholar. When he began teaching in his early twenties, he soon began to attract many disciples, helping them solve problems. During his life about 3,000 students, from many other countries like Qi, Chu, Wei and Jin, came a long way to have him as their teacher. His students reached great achievements. In virtues, Yan Yuan and Zhong Gong were the prominent ones; in language study, Zai Wo and Zi Gong were outstanding, and in politics, Ran You and Zi Lu showed great talent. He not only created some effective teaching methods, but also proper studying techniques. For the rest of his life, fearing that it would be lost during the troubled era in which he lived, he put in order some ancient literature and edited *The Spring and Autumn Annals* (the first Chinese historical records). It is generally believed that Confucius preserved the works of Chinese literature at that time for later generations. He also left *The Analects* for posterity, a collection of his thoughts and teachings, compiled by his disciples in memory of Confucius.

2. His Political Career

He was not only a scholar, but also a man who had political ideals. Though his educational career was fruitful, his political ideas were not accepted by the rulers. He only served the government for four years due to disagreements with the ruler. During the following fourteen years, he left his country and traveled around other countries to present his ideas to different rulers. Finally, because his ideas were not adopted, he returned home, still not recognized by the king.

3. His Ideas

Confucius was born only a few years after the Buddha. Unlike the Buddha, however, he did not seek to escape from the world, but wanted to find a way for man to be happy on earth. His main idea is to administer the country with morals. Regarding personal relationships he once said that "Do not do unto others what you would not want others to do unto you." He believed that the ruler was like the father in a family: he directed the government, but was responsible for the welfare of his people. He taught that human nature was good, rather than bad. If men would think and act properly, he believed, most evils would disappear. His teachings held that men should develop the virtues of kindliness, tolerance, and respect for older people and ancestors. He stressed the importance of education, good manners, and tradition.

It was his belief that education should not be the privilege of a limited number of people. So he was the first person to bring the knowledge previously reserved for the ruling class to the common people. His thoughts and teachings were highly valued by the Chinese people and they formed the foundation of a great philosophy, Confucianism.

In short, he aimed to establish a world of great harmony. For over two thousand years, Confucianism has guided numerous people's behavior and has been the mainstream of Chinese culture. In recent years, his great ideas have been accepted by many people all over the world. (656 words)

Task 2 **Let's Learn**

1 Read the following words and expressions, and tick the ones you know in the last column of the word list.

Numbers	Words & Expressions	Meanings	Notes
1	descendant /dɪˈsendənt/	*n.* 祖孙后代，后裔	
2	royal /ˈrɔɪəl/	*adj.* 王室的，高贵的	
3	poverty /ˈpɒvəti/	*n.* 贫困，缺乏	
4	financial /faɪˈnænʃl/	*adj.* 金融的，财经的	
5	menial /ˈmiːniəl/	*adj.* 低贱的，不需要技能的	
6	livestock /ˈlaɪvstɒk/	*n.* 家畜，牲畜	
7	insatiable /ɪnˈseɪʃəbl/	*adj.* 贪得无厌的，无法满足的	
8	disciple /dɪˈsaɪpl/	*n.* 门徒，弟子	
9	previously /ˈpriːviəsli/	*adv.* 先前，预先	
10	reserve /rɪˈzɜːv/	*v.* 预订（座位等）；储备；拥有	
11	foundation /faʊnˈdeɪʃn/	*n.* 基础；地基；基金会	
12	edit /ˈedɪt/	*vt.* 编辑；校订	
13	preserve /prɪˈzɜːv/	*vt.* 保存；保护；维持	
14	tolerance /ˈtɒlərəns/	*n.* 公差；宽容；容忍	
15	ancestor /ˈænsestə(r)/	*n.* 始祖，祖先	
16	welfare /ˈwelfeə(r)/	*n.* 福利；幸福	
17	privilege /ˈprɪvəlɪdʒ/	*n.* 特权；优待	
18	influence /ˈɪnfluəns/	*n.* 影响 *vt.* 影响；改变	
19	probably /ˈprɒbəbli/	*adv.* 大概；或许；很可能	
20	analects /ˈænəlekts/	*n.* 文选；论集	
21	Buddha /ˈbʊdə/	*n.* 佛陀；佛像	

2 Read and learn the key terms of Chinese philosophy ideas and doctrines, and tick the ones you know in the last column of the word list.

Numbers	Terms	Meanings	Notes
1	Confucianism	儒家思想/学说	
2	Confucius/Mencius/Xun Zi	孔子/孟子/荀子	
3	The Four Books and the Five Classics	四书五经	
4	Three Thousand Disciples of Confucius	孔子弟子三千	
5	to make no social distinctions in teaching	有教无类	

Numbers	Terms	Meanings	Notes
6	to teach students in accordance of their aptitude	因材施教	
7	teaching methods	教学方法	
8	the eliciting method	诱导法	
9	to combine theory with practice	理论与实践相结合	
10	to express one's own views	发表观点	
11	to analyze problems	分析问题	
12	to solve problems	解决问题	
13	the goodness of human nature	人性本善	
14	the virtues of kindliness and tolerance	仁义与宽恕	
15	the virtues of respecting for the old and ancestor	尊敬长辈	
16	Li or etiquette	礼仪	
17	benevolent government	仁政	
18	People are more important than the ruler.	民贵君轻	
19	Taoism	道家学说	
20	Lao Zi/Zhuang Zi	老子/庄子	
21	Tao follows nature.	道法自然	
22	to achieve the harmony between human and nature	天人合一	
23	to conform to the nature	顺应天道	
24	to transform between Yin and Yang	阴阳互转	
25	to take no action or take actions	无为无不为	
26	*Tao Te Ching/Daodejing*	《道德经》	
27	*Peripateticism*	《逍遥游》	
28	Zhuangzhou's Dreaming of Butterfly	庄周梦蝶	
29	Mohism	墨家学说	
30	Mo Zi	墨子	
31	universal love, opposing war, frugality	兼爱、非攻、节用	
32	legalism	法家学说	
33	Han Feizi	韩非子	
34	revolution	变革	
35	to enrich the country and empower the army	富国强兵	
36	rule by law	以法治国	
37	military strategist in ancient China	兵家思想	
38	*Art of War*	《孙子兵法》	

3 Get to know some useful expressions for etiquette of Chinese philosophy.

(1) 三人行，必有我师焉。择其善者而从之，其不善者而改之。

If three of us are walking together, at least one of the other two is good enough to be my teacher. Choose the good to follow, the bad to change.

(2) 朽木不可雕也。

One cannot carve in rotten wood./Decayed wood cannot be carved./A useless fellow.

(3) 君子坦荡荡，小人长戚戚。

A gentleman is open and poised, but a villain is mean and swayed.

(4) 敏而好学，不耻下问。

Being intelligent, as forbidden fruit is sweet.

(5) 学而不思则罔，思而不学则殆。

Learning without thought is lost; thought without learning is idleness.

(6) 不闻不若闻之，闻之不若见之，见之不若知之，知之不若行之。学至于行而止矣。行之，明也。

Tell me, I forgot; Teach me, I remembered; Involved me, and I learned.

(7) 道生一，一生二，二生三，三生万物。

The Tao gives birth to One. One gives birth to Two. Two gives birth to Three. Three gives birth to all things./The Tao or Dao is the birth of everything.

(8) 人法地，地法天，天法道，道法自然。

Man follows the law of the earth, the earth the law of the heaven, the heaven the law of the Tao, the Tao the law of nature.

(9) 上善若水，水善利万物而不争。

The highest excellence is like (that of) water. The excellence of water appears in its benefiting all things.

(10) 欲速则不达。

Haste makes waste./More haste, less speed.

Task 3 Consolidated Exercises

1 Decide whether the statements are True or False according to the passage.

(1) Confucius's father left his son and his wife a royal rich family after his death.

(2) His political ideals were accepted by the rulers in his time.

(3) People think that it is Confucius that saved the works of Chinese literature at that time for later generations, and he left The Five Classics and *The Analects*.

(4) *The Analects* is a collection of Confucius's thought and teaching ideas.

(5) According to paragraph 4, we can infer that Confucius and Buddha, who escaped from the world, were born at the same era.

(6) One of his core ideas of teaching is the goodness of human nature.

(7) His political idea is to be cruel to the people.

(8) From paragraph 5, we can know that education belonged to the ruling class before Confucius.

(9) The concepts of Confucius only had a great impact on the society at his time.

(10) The passage mainly talked about what life Confucius lived in his lifetime.

2　Fill in the blanks with the proper words given and change the word form if necessary, or choose the best choices from the blanks.

(1) He became _____ (know) as a great man after studying for his whole life.

(2) Because of the spread of COVID-19 in the year of 2020, many people have few chance to _____ (take on) a new job.

(3) Confucius _____ (devote) his whole life to the development of education.

(4) The man spent a week _____ (collect) materials for his essay.

(5) After coming back from his journey to the western of China, he made up his _____ (heart, mind, face) to work hard to help those in poverty.

(6) Those who are thirst _____ (of, on, for) learning will be the winner in the future, to some extent.

(7) _____ (Respect, Respecting, Respected) others is a virtue that Confucius taught in his teaching.

(8) As college students, we should learn to _____ (put, putting, putted) what we learned into practice.

3　Translate the following sentences into English.

(1)现今人们相信地球是圆的。（It is generally believed that...）

(2)他是在北京的时候遇到了他最好的老师。（It is... that...）

(3) 人之初,性本善。

(4) 道可道,道法自然。不敢为天下先。

(5) 故明主之治国也,明赏,则民劝功;严刑,则民亲法。

Section III
Let's Practice

Task 1 Let's Talk

1 Watch the video to know some information of another great philosopher, Lao Tzu, and learn to introduce a person briefly.

Kong Qiu (Confucius), the historical Chinese saint, is considered to be the greatest teacher of China. His teacher was Lao Tzu (Lao Zi). Kong Qiu said that Lao Tzu was a dragon who hid in the clouds. He learned from Lao Tzu all his life, but couldn't understand all his great wisdom. 2,500 years later, his work has been translated into many foreign languages. Lao Tzu believed that people should learn from nature and keep a harmonious relationship with nature. Beside a river, Lao Tzu told Kong Qiu that a saint should act like a flowing water. Water gives life to others, but does not fight with others. A saint makes no error and no one hates him. Therefore, he can make great achievements. In China, Lao Tzu is considered as the elite of teachers.

2 Read and repeat the following dialogue.

A: Hello, John. Long time no see. How is it going?

B: Not too bad. Thanks.

A: What have you been busy doing?

B: I've been studying Chinese philosophy. It is amazing and full of mysterious.

A: Yes, we Chinese have 5,000-year-long civilization with some great philosophers and their ideas which have influence on people's life all through the history. We are all proud of them. What do you know about these ancient Chinese philosophers?

B: I know Confucius, Lao Zi, and Zhuang Zi. It's not difficult to understand the ideas of Confucius, but I cannot understand some of the Taoism's doctrine better.

A: Lao Zi is the founder of Taoism. One of his core doctrines is to follow the nature, because human in the nature is humble. So he put forward another idea: wu wei.

B: Does it mean doing nothing?

A: Yes. Just follow the course of Tao/nature, action without intention.

B: So it is called wu wei or wu suowei.

A: Confucianism lays stress on the importance of kindness and rules, like respecting others, courtesy etc., in the society, while Taoism on the naturalness. Both focus on the achievement of harmony between the human and nature. These thoughts have survived many vicissitudes.

B: Quite profound. I need more time to further understand them.

Task 2 | **Team Work**

1 Group discussion and presentation.

Each group chooses one of the doctrines listed on page 10 to discuss and then makes a presentation in class. When discussing, take the following points into account:

(1) What's the origin of the doctrine?

(2) How do you understand this doctrine?

(3) What can you do with the doctrine in your daily life?

2 Critical thinking.

How can you understand the Chinese philosophy "Doctrine of the Mean"? And how does it influence people's life? Give an example to support your ideas.

Section IV
Extended Reading

Taoism

Chinese name—道教/道家思想; literal meaning—"teaching the way". Universally known as Taoism or Daoism.

It is a religious or philosophical tradition of Chinese origin which emphasizes living in harmony with the Tao (Chinese: 道; pinyin: dào; literally: "the Way", also romanized as Dao).

Taoism differs from Confucianism by not emphasizing rigid rituals and social order.

Taoist ethics vary depending on the particular school, but in general tend to emphasize wu wei 无为 (let things take their own course—action without intention), "naturalness", simplicity, spontaneity, and the Three Treasures:

慈 cí "compassion"

俭 jiǎn "frugality"

不敢为天下先 bù gǎn wèi tiān xià xiān "humility"

The founder of Taoism is Lao Tzu—born likely in 6th or 4th Century BC. According to the legend, he left China for the West on a water buffalo. Although he was dressed as a farmer, the border official recognized him and asked him to write down his wisdom. What he wrote became the sacred text known as the *Tao Te Ching* or *Daodejing*. It gives instructions—often vague and open to multiple interpretations—on how to live a good life.

After writing this piece, Lao Tzu is said to cross the border and disappear from the history, perhaps to become a hermit. In reality, the *Tao Te Ching* is likely to be the compilation of the works of many authors overtime.

Dao (Way, Path)

Daoism (Taoism)

- Daoism was based on the ideas and attitudes of Lao Tzu.
- The book called *Tao Te Ching*.
- Everything in nature has two balancing forces called Yin and Yang.
- Dark and light, cold and hot, male and female.
- These forces are always equal and balanced.

Some Enlightenment of Lao Tzu's Wisdom:

1. We ought to make more time for stillness. "To the mind that is still, the whole universe surrenders." We need to let go our schedules, worries and complex thoughts for a while, and simply experience the world. We spend so much time rushing from one place to the next, but Lao Tzu reminds "Nature does not hurry, yet everything is accomplished". Certain things—grieving, growing wiser, developing a new relationship—only happen on their own schedule, like the changing of leaves in the autumn.

2. When we are still and patient, we also need to be open. Lao Tzu says: "The usefulness of the pot comes from its emptiness. Empty yourself of everything, let your mind become still." If we are too busy and too preoccupied with anxiety or human ambition, we will miss a thousand moments of human experience that are natural inheritance. We need to be awake to the sounds of the birds in the morning, the way other people look when they're laughing, the feeling of wind against our face—these experiences re-connect us to the parts of ourselves.

3. "When I let go of who I am, I become what I might be." We need to be in touch with our real selves. We spend a great deal of time worrying who we ought to become. But we should, instead, take time to be who we already are at heart. We might re-discover a generous impulse or a playful side we've forgotten. Our ego is often in the way of our true self, which must be found by being receptive to the outside world rather than focusing on some critical, too ambitious, and internal image.

4. Nature is particularly useful in helping us to find ourselves. Lao Tzu liked to compare different parts of nature to different virtues. "The best people are like water, which benefits all things and does not compete with them. It stays in lowly places that others reject. That is why it is so similar to the Dao." Each part of nature can remind us of the quality we admire and should cultivate ourselves—the strength of the mountains, the resilience of trees, and the cheerfulness of flowers.

(644 words)

Task 1 New Words and Expressions

Read the following words and expressions, and tick the ones you know in the last column of the word list.

Numbers	Words & Expressions	Meanings	Notes
1	ethics /ˈeθɪks/	n. 伦理学; 道德标准	
2	buffalo /ˈbʌfələʊ/	n. 水牛	
3	compassion /kəmˈpæʃn/	n. 同情, 怜悯	
4	resilience /rɪˈzɪliəns/	n. 恢复力; 弹力; 顺应力	
5	frugal /ˈfruːgl/	adj. 节俭的; 朴素的; 花钱少的	
6	preoccupy /priˈɒkjupaɪ/	v. (使) 全神贯注; 提前占据	
7	spontaneity /ˌspɒntəˈneɪəti/	n. 自发性; 自然发生	
8	hermit /ˈhɜːmɪt/	n. (尤指宗教原因的) 隐士	
9	humility /hjuːˈmɪləti/	n. 谦卑, 谦逊	
10	receptive /rɪˈseptɪv/	adj. 善于接受的; 能容纳的	

Task 2 Comprehension

1 After reading the passage, try to retell a story of Lao Tzu in your own words.

_____

_____

_____

2 Read the doctrines of Lao Tzu again, translate them into Chinese, and then discuss them.

(1) To the mind that is still, the whole universe surrenders.

(2) Nature does not hurry, yet everything is accomplished.

(3) The usefulness of the pot comes from its emptiness. Empty yourself of everything, let your mind become still.

(4) When I let go of who I am, I become what I might be.

(5) The best people are like water, which benefits all things and does not compete with them. It stays in lowly places that others reject. That is why it is so similar to the Dao.

3 Critical Thinking.

List the similarities and differences between Confucianism and Taoism.

Insight into Proverbs

"Ren" 仁 —Benevolence

Benevolence is the common virtue and constant virtue of the Chinese nation. Benevolence refers to the relationship between people and the love of people, from the love of parents, brothers and sisters, and then to the love of others—fraternity of mankind.

1. 与人方便，自己方便。

If you go easy on others, they'll go easy on you.

(If you aren't overly critical of the behavior of others, they will be more likely to not give you a hard time.)

2. 得放手时须放手；得饶人处且饶人。

When you should let others off the hook, you must let them off the hook; when you should pardon others, you had best pardon them.

(Be lenient whenever possible, with the implication that others then will tend to show you leniency.)

3. 己所不欲勿施于人。

What you do not want others to do unto you, do not do unto them.

4. 在家不行善，出门大雨淋。

If at home you do not act morally, you'll encounter a storm when you go out your door.

(If in our daily lives we do immoral things, we'll eventually meet with disaster out in the world.)

5. 仁爱必须由近及远。

Charity begins at home, but should not end there.

6. 仁者不忧。

Moral men are free from anxiety.

7. 养气以立德。

Nurturing qi(integrity) is to cultivate morality.

8. 仁者，人也，亲亲为大。

The greatest love for people is the love for parents.

9. 德不孤，必有邻。

A man of virture will not be isolated.

10. 安慰胜过傲慢。

Comfort is better than pride.

11. 君子以文会友，以友辅仁。

A wise man makes friends by his taste for art and literature. He uses his friends to help him to live a moral life.

12. 刚、毅、木、讷，近仁也。

Unbending strength, resoluteness, simplicity and slowness of speech are close to benevolence.

13. 仁者爱人。

The moral life of a man consists in loving men.

14. 有德者必有言，仁者必有勇。

A man who possesses moral worth will always have

something to say worth listening to.

(A moral character always has courage.)

15. 民之于仁也，甚于水火。

Men need morality more than the necessaries of life,

such as fire and water.

Section VI
Self-assessment Checklist

1 Now, it's time for you to review your performance after learning this unit. Carry out a self-assessment by checking the following table.

Items	Ratings			
1. Knowledge	A	B	C	D
I know the representative figures of ancient Chinese philosophy.				
I know the books and doctrines of different schools of ancient Chinese philosophy.				
I master the ancient Chinese philosophy's influence on the Chinese.				
I know the common principles of ancient Chinese philosophy.				
2. Skills	A	B	C	D
I can guess the general idea of a paragraph by using topic sentence.				
I can identify the appositive structure to help understand the reading.				
I can use synonyms to expand my vocabulary.				
3. Speaking	A	B	C	D
I can talk one of the ancient Chinese philosopher fluently.				
I can explain one of the doctrines of Chinese philosophy logically.				
4. Confidence in Chinese Culture	A	B	C	D
I can understand the essence of ancient Chinese philosophy.				
I have the awareness of integrating ancient Chinese philosophy with western one.				
I feel proud of ancient Chinese philosophy's influence in the world.				

A: Basically agree

B: Agree

C: Strongly agree

D: Disagree

2 Fill in the blanks in the mind maps below to check whether you have a good understanding of this unit.

(1) Outlines of ancient Chinese philosophy.

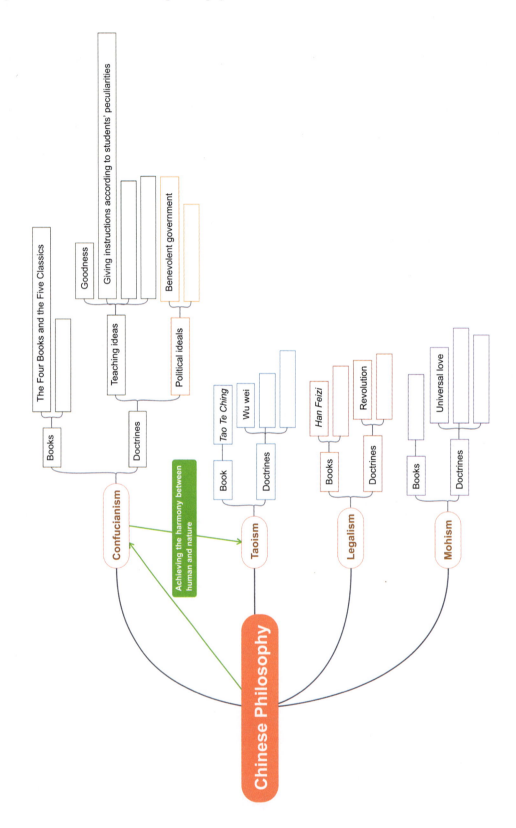

(2) Language points related to this unit.

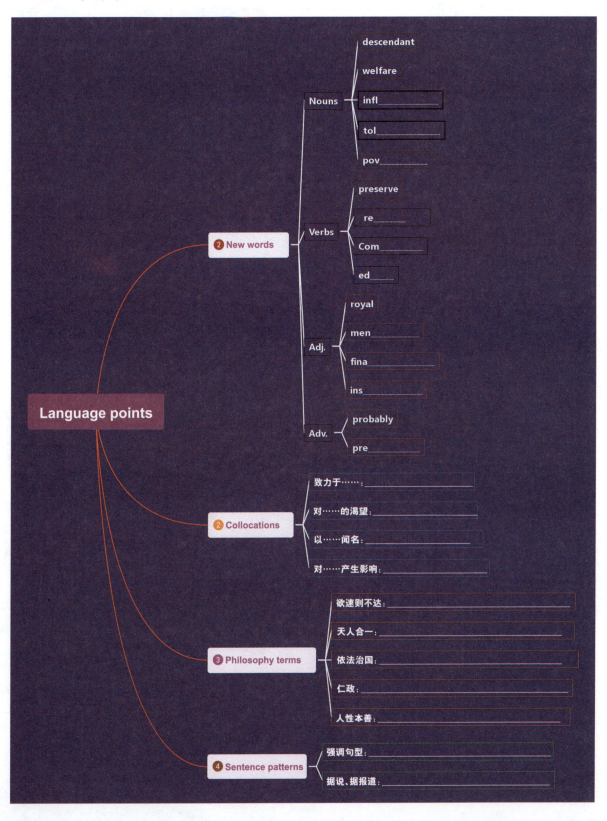

(3) Write down one of your favorite expressions of ancient Chinese philosophy, and state your reasons briefly in a logical way.

_____

_____

_____

_____

_____

_____

_____

_____

_____

_____

_____

_____

_____

_____

_____

_____

_____

_____

_____

_____

_____

_____

_____

Unit 2

Chinese Zodiac and Solar Terms

Do you know why Chinese people have 12 Zodiacs and what does each zodiac mean? Why do Chinese people think it important to wear red in his/her year? Do you ever think how ancient Chinese people predict the weather? Why disease prevention is the core ideal of traditional Chinese medicine for people to stay health? What does the saying "Eating carrot in Winter and ginger in Summer" mean? All is related to Chinese calendar that is divided into lunar and solar, taking into account the longest and the shortest days of the year. These particular features involved in all aspects of ancient Chinese people have greatly influenced people's life from generation to generation. Even in modern times, Chinese people still rely on these ancient methods to keep order in their busy and chaotic lives. Let's scan the overview to know what we will learn in this unit.

Overview

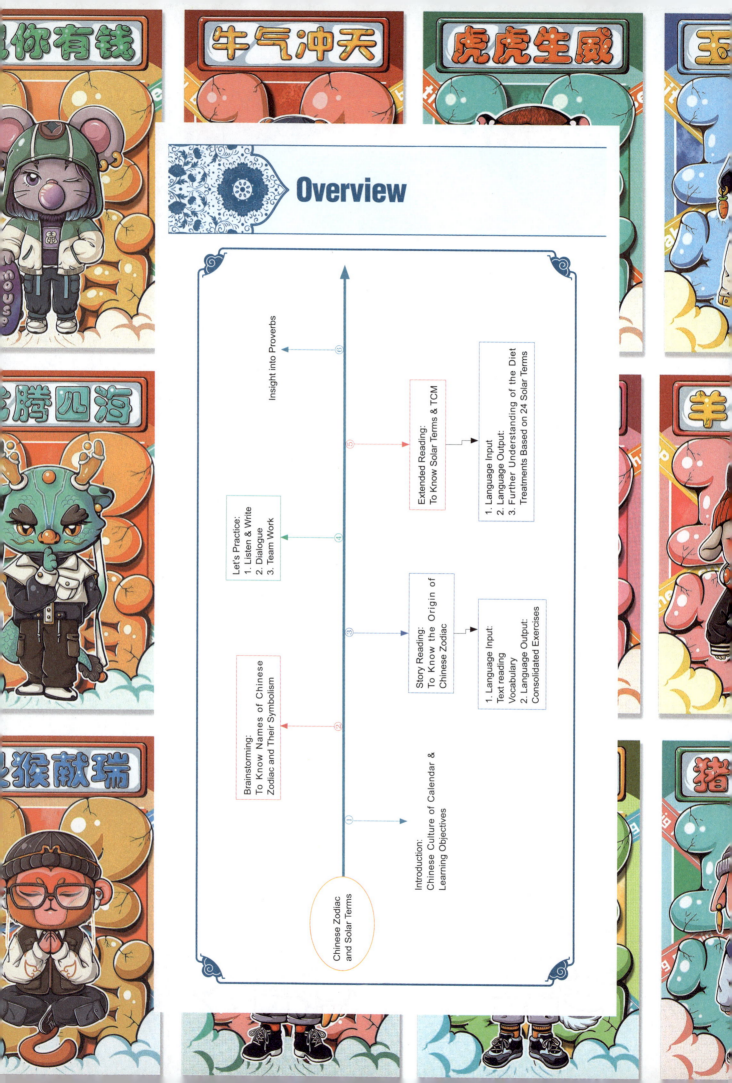

Chinese Zodiac and Solar Terms

① Introduction:
Chinese Culture of Calendar & Learning Objectives

② Brainstorming:
To Know Names of Chinese Zodiac and Their Symbolism

③ Story Reading:
To Know the Origin of Chinese Zodiac

1. Language Input:
Text reading
Vocabulary
2. Language Output:
Consolidated Exercises

④ Let's Practice:
1. Listen & Write
2. Dialogue
3. Team Work

⑤ Extended Reading:
To Know Solar Terms & TCM

1. Language Input
2. Language Output:
3. Further Understanding of the Diet Treatments Based on 24 Solar Terms

⑥ Insight into Proverbs

Learning Objectives

After learning this unit, students are able to reach the goals below.

1	专业能力目标	① 了解中国民俗文化——十二生肖、二十四节气、中医的相关知识及其对人们生活的影响
		② 扩大与十二生肖、二十四节气、中医相关的词汇量
		③ 提升阅读速度,快速掌握文章的主旨和大意;了解十二生肖的文化内涵和中国夏历年的构成;用较流利的语言传播中国远古时期国人根据星象变化而形成的"天人合一"的思想
		④ 了解二十四节气对中医食疗的影响及中医疗法的原理;能运用恰当的语言讲述古人的天文学智慧
2	方法能力目标	① 辨识主题句,学会写出简洁的主题句,学会篇章布局
		② 正确使用动名词做主语的表达方式;提高对句子复杂性的认识
		③ 理解独立主格结构的功能并能在句子中使用
		④ 学会灵活使用标点符号,增加句式表达的多样性
3	社会能力目标	① 通过主题阅读,拓展知识面,丰富人生阅历,提升社交能力
		② 在信息技术时代,具备获取与天文学、星相学和中医养生相关信息的能力和分析相关信息的能力
		③ 掌握基本阅读技巧,提高学习效率和社会工作效率
		④ 通过主题阅读,了解古人天文学和人体脉络相结合的知识;学会恰当地运用中医国粹智慧调养生息,提升生活质量
4	情感与思政目标	① 通过本单元的学习,提高对中医食疗的判断能力和决策能力;传承中医文化的精髓;弘扬爱国情怀
		② 通过主题阅读,感受中医的博大精深,加强文化自信
		③ 通过主题阅读及拓展阅读,提升合理运用二十四节气进行食疗的能力;辩证地看待中医和西医的关系;弘扬中华民族优秀传统文化

Section I
Brainstorming

Task 1 | Let's Warm Up

Look at the following pictures, write down the signs of animals below and put the number before their names according to the order of the Chinese Zodiac.

Task 2   Let's Watch and Say

1 Watch the video *Chinese 12 Animal Zodiacs* and try to write down the symbolism of each animal sign in Chinese culture.

Rats are _____.

Oxen are _____.

Tigers are _____.

Rabbits are _____.

Dragons are _____.

Snakes are _____.

Horses are _____.

Goats are _____.

Monkeys are _____.

Roosters are _____.

Dogs are _____.

Pigs are _____.

2 Critical thinking.

(1) Why are there no lion and cat in Chinese Zodiac?

(2) As it's impolite and rude to talk about personal affairs, like age and salary, to avoid the dilemma, what is the proper way to communicate with others, especially a foreigner? If you are asked to make a suggestion to your foreign client about gifts related to Chinese Zodiac to his 12-year-old son, what would you say?

_____

_____

_____

_____

_____

_____

_____

_____

Section II
Story Reading

Task 1   Let's Read

Chinese Zodiac

1. Legends and mythology are a big part of Chinese culture, especially in relation to the Chinese Zodiac. The 12 animals include Rat, Ox, Tiger, Rabbit, Dragon, Snake, Horse，Goat, Monkey，Rooster, Dog and Pig. It's an unusual combination of animals for sure, and their appearance on the Chinese Zodiac is the topic of countless legends and is deeply embedded in Chinese mythology. Along with birth year animals, the Chinese Zodiac also represents inner animals and secret animals. We're probably all aware of the 12 animals or signs depicted on the Chinese Zodiac. Based on our year of birth, we understand that we're born under the sign of the Dragon, or the tiger, or one of the 10 other signs. But the truth is it's the inner animals and the secret animals that actually tell the most about who we really are and what our futures hold.

2. "Sheng" in Chinese is "birth" and "xiao" is "resemblance". To discuss about the Shengxiao of Chinese, it is better to say something about the chronological years recorded by Stems and Branches in the Chinese lunar calendar. As we know, the lunar calendar was already perfect over 3,000 years ago during Xia Dynasty, so it was also called "Xia Calendar". In Xia Calendar, the chronological years were recorded on the basis of Stems and Branches, that is on combining in succession one Heavenly Stem and one Earthly

Branch. The Heavenly Stems are 10, namely: *Jia*, *Yi*, *Bing*, *Ding*, *Wu*, *Ji*, *Geng*, *Xin*, *Ren* and *Gui*. The Earthly Branches are 12, that is: *Zi*, *Chou*, *Yin*, *Mao*, *Chen*, *Si*, *Wu*, *Wei*, *Shen*, *You*, *Xu* and *Hai*.

3. Combining one Heavenly Stem with one Earthly Branch in succession will give *Jiazi*, *Yichou*, *Bingyin*… and so on to the sixtieth one—*Guihai* to make up a set, which is generally known as "Sixty Jiazi". These are then matched to the chronological years, with 60 years as a cycle, and similar to years, months and days, and all have their own Stems and Branches, so the records in the calendar on this basis would be very clear and in a good order.

4. The Chinese ancestors emphasized the unity of heaven and man. Human beings and all life in the nature were thought to be naturally associated, the universe was co-existing with us, and everything around was combined with us as unity. Hence, 12 animals were used to match 12 Earthly Branches, such as *Zi* is Rat, *Chou* is Ox… In calculating years, for example, any year with "*Chen*", such as *Jiachen*, *Bingchen* and so on, will be called the dragon year, and those born in the dragon years will belong to the Dragon. Later, the year one is born is thought to be related to his behavior, which will resemble that of the animal of the Earthly Branches. For example, those born in dragon years will resemble dragon.

5. The 12 Shengxiao are *Zi* Rat, *Chou* Ox, *Yin* Tiger, *Mao* Rabbit, *Chen* Dragon, *Si* Snake, *Wu* Horse, *Wei* Goat, *Shen* Monkey, *You* Rooster, *Xu* Dog and *Hai* Pig. (524 words)

Task 2　Let's Learn

1 Read the following words and expressions, and tick the ones you know in the last column of the word list.

Numbers	Words & Expressions	Meanings	Notes
1	legend /ˈledʒənd/	*n.* 传奇；说明	
2	mythology /mɪˈθɒlədʒi/	*n.* 神话；神话学	
3	calendar /ˈkælɪndə(r)/	*n.* 日历；[天] 历法	
4	combination /ˌkɒmbɪˈneɪʃn/	*n.* 结合；组合	
5	combine /kəmˈbaɪn/	*vt.* 使化合；使联合 *vi.* 联合，结合	
6	embed /ɪmˈbed/	*vt.* 栽种；使嵌入	
7	depict /dɪˈpɪkt/	*vt.* 描述；描画	
8	resemblance /rɪˈzembləns/	*n.* 相似；相似之处	
9	resemble /rɪˈzembl/	*vt.* 类似，像	

Numbers	Words & Expressions	Meanings	Notes
10	chronological /ˌkrɒnəˈlɒdʒɪkl/	*adj.* 按发生时间顺序排列的，按时间计算的	
11	lunar /ˈluːnə(r)/	*adj.* 月亮的，月球的；阴历的	
12	dynasty /ˈdɪnəsti/	*n.* 王朝，朝代	
13	succession /səkˈseʃn/	*n.* 连续；继位；继承权	
14	emphasize /ˈemfəsaɪz/	*vt.* 强调	
15	associate /əˈsəʊsieɪt; əˈsəʊʃieɪt/	*v.* 联想，联系	
16	co-exist /kəʊ ɪɡˈzɪst/	共存	
17	universe /ˈjuːnɪvɜːs/	*n.* 宇宙，世界	
18	unity /ˈjuːnəti/	*n.* 团结；一致	
19	calculate /ˈkælkjuleɪt/	*vi.* 计算；以为　*vt.* 计算；预测	
20	hence /hens/	*adv.* 因此；今后	
21	unusual /ʌnˈjuːʒuəl; ʌnˈjuːʒəl/	*adj.* 不寻常的；与众不同的	
22	in relation to	涉及，关于	
23	be embedded in sth.	植根于……	
24	along with	顺着……；和……一道（一起）；连同	
25	be matched to	匹配	
26	be aware of	意识到，知道	
27	to make up a set	组成一组	
28	belong to	属于，归属	

2 Read and learn the key terms of Chinese Zodiac, and tick the ones you know in the last column of the word list.

Numbers	Terms	Meanings	Notes
1	Chinese Zodiac	中国生肖	
2	Shengxiao	生肖	
3	lunar calendar	阴历	
4	the Heavenly Stem and Earthly Branch	天干地支	
5	Heavenly Stem	天干	
6	Earthly Branch	地支	
7	Sixty Jiazi	60甲子	
8	the unity of heaven and man	天人合一	
9	one's birth year, one's animal year, one's year	本命年	

新发展中国传统文化经典教程

3 Get to know some useful expressions for etiquette of Chinese Zodiac.

Asking someone's age may be impolite and rude, so how to know his or her age without offending the etiquette is crucial to communication both in daily life and business. After learning this unit, we can apply Chinese Zodiac in our communication with others socially, at the same time we can promote Chinese culture. So we should learn to use the following expressions for etiquette of Chinese Zodiac:

What is your animal sign?

What animal do you belong to?

What's your Chinese Zodiac/Shengxiao?

This is my year.

Task 3   Consolidated Exercises

1 Answer the following questions briefly according to the passage.

(1) What do inner animals and secret animals in Chinese Zodiac actually symbolize?

(2) What does "sheng" and "xiao" represent in Chinese respectively?

(3) When did the Lunar Calendar come into being?

(4) How did the Lunar Calendar calculate the year?

(5) How many Heavenly Stems and Earthly Branches are there in Chinese Lunar Calendar?

2 Make a mind map of the heavenly stems and the earthly branches and mark the names with both pinyin and Chinese characters.

3 Fill in the blanks with the proper words given, and change the word form if necessary.

legend	combine	embed	emphasise
universe	resemble	along with	belong to

(1) To many, a _____ must be a story, with characters and some sort of plot.

(2) He said he would come _____ his family to China.

(3) Many cultures around the world are deeply _____ in the mythology.

(4) The construction of Xi'an city _____ that of Beijing.

(5) We should learn to _____ Chinese culture with our daily life.

(6) Cats, dogs, and oxen _____ the livestock.

(7) We have an opportunity now to really unlock the secrets of the _____.

(8) To _____ the point, Mr. Ahern has promised to appoint Mr. Reynolds as his personal envoy (使节) to Northern Ireland.

4 Translate the following sentences into English with the terms or expressions learned in this passage.

(1) 增强文化意识和文化自信是我党对当代大学生提出的一项培养目标。

(2) 中国古人强调天人合一的理念。

(3) 人类与宇宙共存, 世间万物融合为一个整体。

(4) 这种理念深植于传统文化。

(5) 你是属什么的?

(6) 中国人认为本命年会犯太岁, 所以都要穿红色的衣服。

(7) 中国的农历年由10个天干和12个地支组成。

(8) 根据中国属相, 不同的动物代表不同的内涵, 如狗代表忠诚、虎代表勇敢。

Section III
Let's Practice

Task 1　Let's Talk

1 Listen to the story and write down the proper answers in the blanks.

It was coming up to the new year. _____(1)_____ animals were arguing who should be the name of the new year. So the Jade Emperor decided to _____(2)_____ the new year name to the winner who first reached the river's other side. So all the animals _____(3)_____. With a mighty splash, the animals leaped into the river and started swimming _____(4)_____ they could towards the other side.

The _____(5)_____ wasn't the best swimmer, but he was the cleverest animal. He saw that the _____(6)_____ was going to win the race, so he swam as fast as he could and grabbed hold of the ox's _____(7)_____. He climbed up the tail onto the ox's back. Just before the ox reached the other _____(8)_____ of the river, the rat leaped over his head onto the bank, taking the _____(9)_____ place, while the ox the second. One by one, the other animals finished the race. Then the Jade Emperor said, "We will _____(10)_____ a year after each of you in the same order that you finished the race." Since then Chinese new year starts from rat.

2 Read and repeat the dialogue.

A: I am a monkey, so this is my year. I'm guessing I'm going to have a pretty lucky year this year.

B: Not so quick, humble. Actually when it's your year it is believed you are going to get more bad luck.

A: Really?

B: Yeah. The Chinese think that this year you are going to offend the God of Taisui.

A: Is there anything I can do to appease the god of age to protect myself?

B: Yeah. There are a few things you can do. You just wear red. Deck yourself out with red.

A: Oh, OK. That's good.

B: In China right now you're seeing so much red. Red is a very auspicious color. I suppose it is really the color of New Year. It is, it's mostly happiness, joy, luck…

Task 2　Team Work

Imitate the above dialogue first, then make one with your partner about your animal sign and perform it in class.

Solar Terms and TCM

Solar terms is a calendar of twenty four periods and climate to govern agricultural arrangements in ancient China and functions even now. As we have mentioned the Chinese calendar is a lunisolar calendar, it takes into account the longest and the shortest days and the two days each year when the length of the day equals that of the night. In other words, the significant days are the Summer and Winter Solstices and the Spring and Autumn Equinoxes.

1. What Are "Jie Qi" and "Zhong Qi"?

Each of these 24 solar terms each suggests the position of the sun every time when it travels 15 degrees on the ecliptic longitude (黄道经度). In each month there are often two solar terms: the first one is generally named "Jie Qi" and the other one "Zhong Qi". Their dates are mirrored by the Gregorian calendar (公历), so we find that during

the first half of a year "Jie Qi" is around the 6th day of a solar month, "Zhong Qi" is around the 21st; in the second half of a year, "Jie Qi" is around the 8th and "Zhong Qi" is around the 23rd.

The 24 solar terms classify the whole year into 24 phases in order in accordance to the earth's climatic change caused by the change of the sun's site change. Life of ancient Chinese people was absolutely in relation to the solar terms, including agricultural arrangements, daily diet and health maintainance, and doing sacrifice.

2. Ways of Maintaining Health in Traditional Chinese Medicine (TCM)

Rain is more and temperature becomes higher. It's time to nurse and harmonize liver and spleen (脾脏), so take more outdoor activities to prevent being spring sleepy.

The rains (Yushui, 雨水) is the period of rainfall in the whole year. In TCM theory, the period of the rains plays the important role in caring spleen and stomach. The harmony of spleen and stomach can improve and regulate the physical metabolism. In TCM, the root of human health is Yuan Qi (元气, Original Qi), and the root of Yuan Qi in human body is spleen and stomach. The poor spleen and stomach are the predominant causes of different diseases. So in the aspect of eating, people had better not eat too much oily and fat food but the red Chinese date, lotus, leek, spinach, orange, honey and sugarcane.

The heart of TCM is the tenet that the root cause of illnesses, not their symptoms, must be treated. And TCM treatments are usually based on the following theories or concepts:

(1) Five-Element Theory. The Five-Element Theory, the bed rock of TCM, systematizes patterns of perceived related phenomena into five major groups named for the universal elements: Woods, Fire, Earth, Metal and Water, which are related major organ systems.

(2) Yin-yang Theory. TCM understands that everything is composed of two complementary energies: one energy is yin and the other is yang. They are never separate; one cannot exist without the other. This is the yin/yang principle of inter-connectedness and interdependence.

(3) Meridian Theory. Meridians, or channels, are invisible pathways through which Qi flows that form an energy network that connects all parts of the body, and the body to the universe. TCM understands that our body has twelve major meridians. Each one is related to a specific organ system.

So disease is believed to be prevented if one follows the health regimen of the 24 solar terms to arrange one's daily diet. (584 words)

Task 1　New Words and Expressions

Read the following words and expressions, and tick the ones you know in the last column of the word list.

Numbers	Words & Expressions	Meanings	Notes
1	solar /ˈsəʊlə(r)/	*adj.* 太阳的；日光的	
2	mention /ˈmenʃn/	*vt.* 提到，谈到	
3	classify /ˈklæsɪfaɪ/	*vt.* 分类；分等	
4	maintain /meɪnˈteɪn/	*vt.* 维持；继续；维修	
5	spleen /spliːn/	*n.* 脾脏	
6	regulate /ˈreɡjuleɪt/	*vt.* 调节，规定；控制	
7	metabolism /məˈtæbəlɪzəm/	*n.* [生理] 新陈代谢	
8	predominant /prɪˈdɒmɪnənt/	*adj.* 主要的；卓越的	
9	lotus /ˈləʊtəs/	*n.* 莲花；荷花	
10	leek /liːk/	*n.* 韭葱	
11	spinach /ˈspɪnɪtʃ; ˈspɪnɪdʒ/	*n.* 菠菜	
12	tenet /ˈtenɪt/	*n.* 原则；信条	
13	symptoms /ˈsɪmptəms/	*n.* [临床] 症状	
14	systematize /ˈsɪstəmətaɪz/	*vt.* 使系统化	
15	perceive /pəˈsiːv/	*v.* 察觉，感觉，感到，感知	
16	phenomena /fəˈnɒmɪnə/	*n.* 现象	
17	compose /kəmˈpəʊz/	*v.* 构成；写作	
18	complementary /ˌkɒmplɪˈmentri/	*adj.* 补足的	
19	meridian /məˈrɪdiən/	*n.* 子午线，经线；中医经脉	
20	invisible /ɪnˈvɪzəbl/	*adj.* 无形的，看不见的	
21	Solar Terms/Jie Qi	节气	
22	lunisolar calendar	阴阳历	
23	the Summer Solstices	夏至	
24	the Winter Solstices	冬至	
25	the Spring Equinoxes	春分	
26	the Autumn Equinoxes	秋分	
27	the ecliptic longitude	黄道经度	
28	the Gregorian calendar	公历	
29	in order	按顺序	
30	in accordance to	根据	
31	health maintainance/regimen	健康养生	
32	Five-element Theory	五行说	
33	Yin-yang Theory	阴阳说	
34	Meridian Theory	经脉说	
35	Traditional Chinese Medicine (TCM)	中医	

Task 2 | **Comprehension**

Answer the questions after finishing reading the passage.

(1) What are the most important days in Chinese solar terms?

(2) How are the 24 solar terms divided?

(3) What are the main causes of all illness according to Traditional Chinese Medicine?

(4) What is the core of Traditional Chinese Medicine?

(5) Which theory do you think is most important for Traditional Chinese Medicine?

Task 3 | **Further Understanding**

Diet Treatments Based on 24 Solar Terms

Feb. 3-5 Spring Begins (立春)

At the beginning of spring, temperature rises and nature wakes. Spring belongs to Wood in the Five Elements, corresponding to liver in the five internal organs, so it's better to take more pungent (辛辣的) food and less acidic (酸的) food to care liver.

Feb. 18-20 The Rains (雨水)

Rain is more and temperature becomes higher. It's time to nurse and harmonize liver and spleen, so take more outdoor activities to prevent being spring sleepy.

Mar. 5-7 Insects Awaken (惊蛰)

Thunder appears and insects awake. Wearing warm clothing, doing exercises and drinking more water help to cope with changeable weather.

Mar. 20-22 Vernal Equinox (春分)

The day and the night half split, or half Yin and half Yang. Be aware to balance Yin and Yang, taking light food to clear body heat and toxic, warm and invigorate Yang Qi.

Apr. 4-6 Pure Brightness (清明)

The sky is clear and the scene on the earth is bright. When Yang rises, tonics (补品) are not recommended. Lite food (低盐饮食、清淡饮食) is to release high blood pressure.

Apr. 19-21 Grain Rain (谷雨)

The temperature rises rapidly and varies greatly, be sure to keep warm and prevent neuralgia (神经痛). More vegetables are to release the internal heat.

May. 5-7 Summer Begins (立夏)

This term belongs to Fire in the Five Element, corresponding to Heart in the Five Internal Organs. Drinking more water is to release internal heat, giving special care for heart.

May. 20-22 Grain Buds (小满)

The weather is humid during this time. Much attention should be paid to prevent measles (风疹). Bland diet (清爽清淡饮食) such as beans, cucumber, lotus root, are recommended with less meat and seafood to eliminate dampness and heat.

Jun. 5-7 Grain in Beard (芒种)

It is humid and hot. Preventing infective disease is important. Enough sleep, nap at noon and sports are recommended. Take self-mental care (调养精神) with lite diet.

Jun. 21-22 Summer Solstice (夏至)

It is the longest day-time and hot weather. It's good time to treat winter diseases (冬病夏治). Pranayama (呼吸控制) and meditation are needed. Take more water and vitamins with necessary salt taking to relieve summer-heat and eliminate heart-fire.

July. 6-8 Slight Heat (小暑)

It's the time of hot weather with storm. As a result, it is easy to get digestive disease for weak function of the stomach. Pay more attention to food hygiene and conserve Yang Qi of heart.

Other Treatment Ways of TCM

Qigong (气功)

Qigong is an energy practice, generally encompassing simple movements and postures. Some Qigong systems also emphasize breathing techniques.

Acupuncture (针灸)

Acupuncture refers to the insertion of needles in acupoints to help Qi flow smoothly.

Herbal Therapy (草药疗法)

Herbal therapy means the use of herbal combinations or formulas to strengthen and support organ system function.

Cupping Therapy (拔火罐)

Cupping therapy is a treatment using a vacuum cup sucked firmly on the skin.

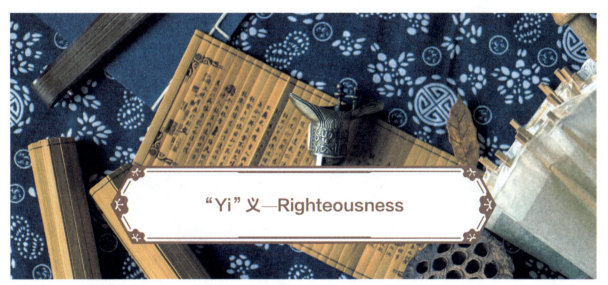

Section V
Insight into Proverbs

"Yi" 义—Righteousness

"Yi" refers to be appropriateness in doing something according to the appropriate time, local conditions, and the people. Do what we ought to do. Confucius said, "The gentleman sees righteousness; the petty man sees profit." So being human, people should have a sense of shame and rightness.

1. 君子义以为质，礼以行之。

A wise and good man makes right the substance of his being; he carries it out with judgment and good sense.

2. 君子上达，小人下达。

A wise and good man looks upwards in his aspirations; a fool looks downwards.

3. 其身正，不令而行。

If a man is in order in his personal conduct, he will get served even without taking the trouble to give order.

4. 秉公办事，何惧天塌下来。

Let righteousness be done, though the heaven falls.

5. 君子成人之美，不成人之恶。

A good and wise man encourages men to develop the good qualities in their nature, and not their bad ones.

6. 不义而富且贵，于我如浮云。

Riches and honours acquired through the sacrifice of what is right, would be to me as unreal as a mirage.

7. 文质彬彬，然后君子。

It is only when the natural qualities and the results of education are properly blended, that we have the truly wise and good man.

8. 树倒猢狲散。

When the tree falls, the monkeys will scatter.
(When an influential person falls from power, his hangers-on disperse.)

9. 见利思义，见危授命。

One who, when he sees a personal advantage, can think of what is right and, in presence of personal danger, is ready to give up his life.

10. 与其责骂罪恶，不如伸张正义。

It is better to fight for justice than to rail at the ill.

11. 过河拆桥

To destroy the bridge after crossing the river
(to pull up the ladder behind them; refers to selfish people who begrudge others from following their path to success by eliminating that path)

12. 多行不义必自毙。

Give a thief enough rope and he'll hang himself.

13. 君子喻于义，小人喻于利。

A wise man sees what is right in a question; a fool, what is advantageous to himself.

14. 鼎力相助。

Put your shoulder to the wheel.

15. 无私者无畏。

The best hearts are always the bravest.

Section VI
Self-assessment Checklist

1 Now, it's time for you to review your performance after learning this unit. Carry out a self-assessment by checking the following table.

Items	Ratings			
1. Knowledge	**A**	**B**	**C**	**D**
I know the origin of Chinese zodiac and its symbolism.				
I know the origin of the 24 solar terms and their roles in Chinese people's life.				
I know the basic communication skills and strategies in asking a person's age.				
I master the expressions of Chinese zodiac and 24 solar terms in English.				
I master the relationship between the 24 solar terms and TCM.				
2. Skills	**A**	**B**	**C**	**D**
I can identify and write concise topic sentences.				
I can use *v.*-ing as the subject to improve the understanding of sentence complexity.				
I can use proper punctuation to increase the variety of sentences.				
3. Speaking	**A**	**B**	**C**	**D**
I can talk Chinese zodiac, the 24 solar terms and TCM treatment fluently.				
I can illustrate Chinese zodiac and the 24 solar terms with useful words and expressions.				
I can explain the theories of TCM logically.				
I can recite some proverbs about Yi.				
I can describe Heavenly Stems and Earthly Branches in Chinese Lunar Calendar.				
4. Confidence in Chinese Culture	**A**	**B**	**C**	**D**
I have the awareness of TCM.				
I feel proud of knowing understanding the essence of Chinese solar terms and TCM.				
I can apply solar terms to adopt diet treatments.				

A: Basically agree

B: Agree

C: Strongly agree

D: Disagree

2 Fill in the blanks in the mind maps below to check whether you have a good understanding of this unit.

(1) Mind map of the symbolism of Chinese zodiac.

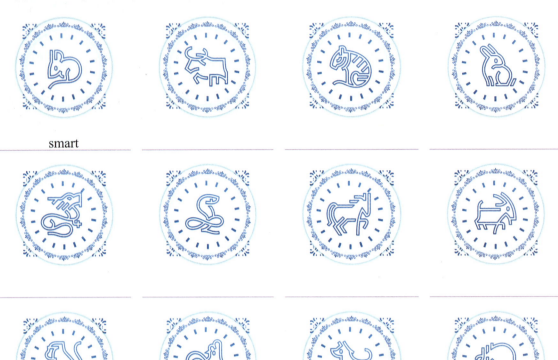

smart

(2) Mind map of 24 solar terms.

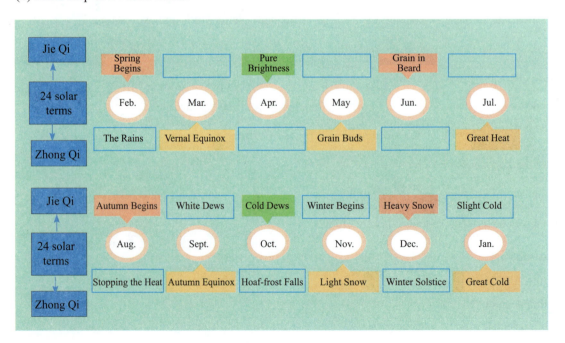

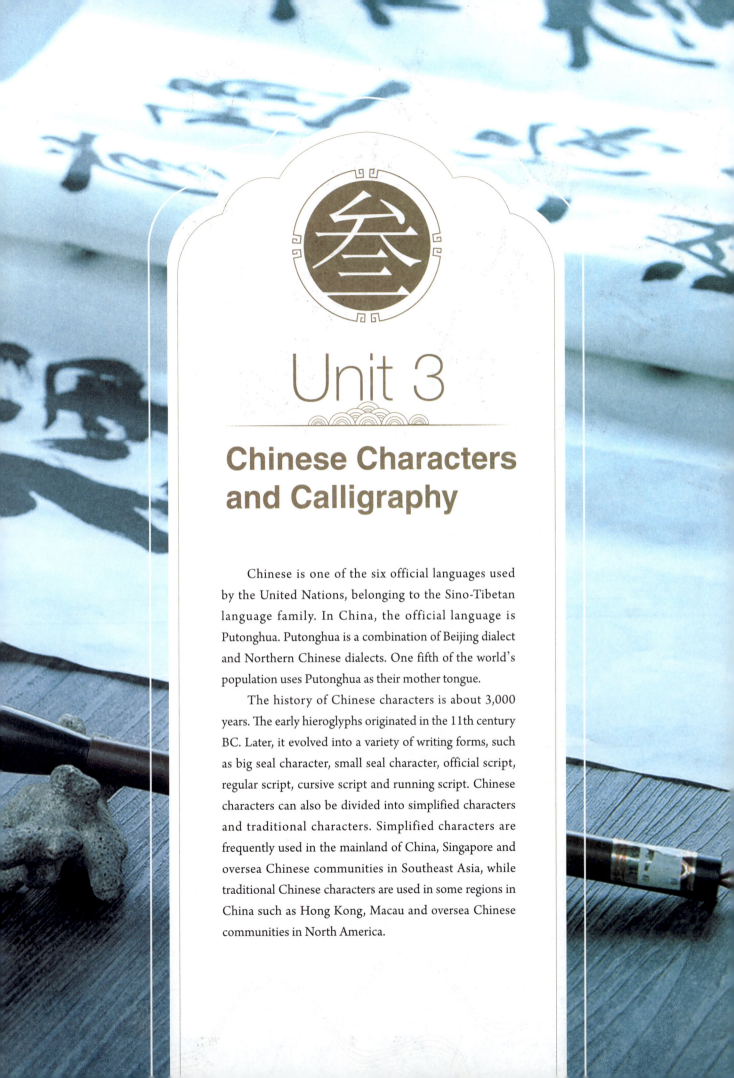

Unit 3

Chinese Characters and Calligraphy

Chinese is one of the six official languages used by the United Nations, belonging to the Sino-Tibetan language family. In China, the official language is Putonghua. Putonghua is a combination of Beijing dialect and Northern Chinese dialects. One fifth of the world's population uses Putonghua as their mother tongue.

The history of Chinese characters is about 3,000 years. The early hieroglyphs originated in the 11th century BC. Later, it evolved into a variety of writing forms, such as big seal character, small seal character, official script, regular script, cursive script and running script. Chinese characters can also be divided into simplified characters and traditional characters. Simplified characters are frequently used in the mainland of China, Singapore and oversea Chinese communities in Southeast Asia, while traditional Chinese characters are used in some regions in China such as Hong Kong, Macau and oversea Chinese communities in North America.

Overview

Chinese Characters and Calligraphy

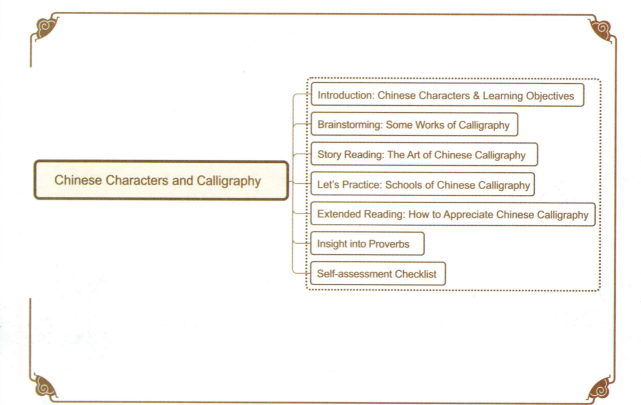

- Introduction: Chinese Characters & Learning Objectives
- Brainstorming: Some Works of Calligraphy
- Story Reading: The Art of Chinese Calligraphy
- Let's Practice: Schools of Chinese Calligraphy
- Extended Reading: How to Appreciate Chinese Calligraphy
- Insight into Proverbs
- Self-assessment Checklist

Learning Objectives

After learning this unit, students are able to reach the goals below.

1	专业能力目标	① 了解书法的起源、演变、发展及其对当今中华文明的塑造与影响
		② 加深对书法艺术相关词汇、句型表达的理解，提高相关词语的表达能力
		③ 提升阅读速度，了解中国书法艺术的鉴赏方法；用较流利的语言概述文章主旨和思想观点；理解并能运用基本的语言技巧及文体修辞；提高阅读理解能力和思想表达能力
2	方法能力目标	① 养成自主阅读的好习惯；根据书法的专题阅读摄取相关信息，扩大知识面
		② 会用略读、寻读、推测等方法摄取书法鉴赏方面的重要信息；能辨识事实与观点；会在阅读中通过构词法、上下文联系等方法猜测单词的意思，扩大词汇量
		③ 会把相关的英语阅读技巧运用于其他学科的学习中，从而提高学习效率
3	社会能力目标	① 通过主题阅读，开拓发散思维与批判性思维；丰富想象力与创新能力
		② 提高获取信息和分析信息的能力
		③ 通过主题学习，提高对汉字书法艺术的认识；提升赏析书法名家作品的能力；能借助书法作品与外商进行有效且有一定深度的沟通与交流
		④ 提高汉字书写水平；提升日常工作中的文字运用能力
4	情感与思政目标	① 通过单元学习，加深对中国书法的了解；对汉字发展的趋势与未来有自身独立的思考；拥有热爱祖国书法文化的情怀；增强文化自信
		② 通过主题阅读，能尊重汉字文化的传统；努力做保护和传承汉字文化的推动者和宣传者
		③ 通过主题阅读及拓展阅读，能运用所学的书法知识解决跨文化交际中的冲突问题；减少文化冲突中的误解，理解文化的包容性

Section I
Brainstorming

Task 1 Let's Warm Up

Match the following calligraphers with their works, and then read the following passage to find out the characters of Chinese calligraphy.

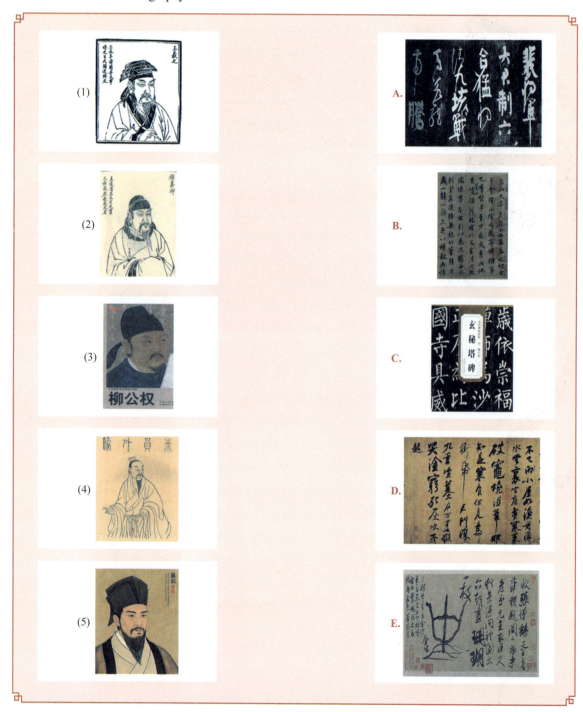

The Characters of Chinese Calligraphy

Chinese calligraphy stresses the overall layout among words and lines. Besides the dense and sparse, proportional arrangement of the strokes of the words, a piece of calligraphy must have an overall harmoniousness that is not lack of changes. Besides the brush, ink and paper are also instruments of calligraphy. Together they produce varied effects. Calligraphists often pay much attention to the methods for the use of ink, which include the use of thick ink, light ink, dry ink, missing ink, wet ink, withered ink, swelling ink, etc.

Task 2 Let's Watch and Say

1 Before watching the video, think about the following questions.

(1) How much do you know about Chinese calligraphy?

(2) What are the basic tools of Chinese calligraphy?

(3) Do you think Chinese calligraphy can reflect the mind of the calligrapher? Why?

2 After watching the video, arrange your mind and make a conclusion of the above questions in pairs.

Section II
Story Reading

Task 1 Let's Read

Chinese Calligraphy

Chinese calligraphy is a unique oriental art form with a brilliant tradition as ancient as the culture itself. All the way through, Chinese calligraphy uses a basic media, brush handling techniques, scripts, presentation and style to express the emotions, culture, artistic and creative feelings, and moral principles of the artist. Birth, development, maturity and flourish of Chinese calligraphy are always closely related to the change of character. For thousands of years' history of Chinese calligraphy, inscriptions on bones or tortoise shells of the Shang Dynasty, dazhuan (large-seal script), xiaozhuan (small-seal script), lishu (official script), kaishu (regular script), xingshu (running script), caoshu (grass script) are all the creations breaking a new way in calligraphy. The present style of Chinese calligraphy is to shape rhythm image with lines. The lines are either strong or soft. The strokes are various. The horizontal and vertical strokes are bend or extensive.

The tip of writing is either dark or clear. Using brush can be slow or quick. Lifting and pressing brush can be both light and weight. The structure is either open or close. The posture is both motive and quiet. The orderly ways are both proportional and new. All the above skills can produce various rhythms. The lines of regular script can make people feel grand and steady. Seal character and official script are both simple and kind. The lines of running script and grass script are much changeable and have thousands of bearings and can show the line's expressive force of rich rhythms and tunes.

The art of Chinese calligraphy has thousands of years' history up to now. There were so many dynasties which make a great Chinese culture. Each dynasty has given birth of many famous calligraphers. Every calligrapher had created a lot of excellent handwriting works in their whole lives. So such tremendous amount of famous calligraphers and excellent works are really culture treasure of our whole nation. The art of Chinese calligraphy with such unique artistic styles is also a brilliant peal in the culture treasury of the whole world .

In the information era, it has been very difficult for Chinese people to know how to solve the problem of Chinese characters' use in computers. People were confronted with the problem of how to type Chinese characters quickly into computers. Through dauntless efforts, various input methods have been invented. With the further application of such technologies as handwriting input and pronunciation input, the ancient Chinese characters have gained a second youth in today's information society and have a more promising future. (423 words)

Task 2 Let's Learn

1 Read the following words and expressions, and tick the ones you know in the last column of the word list.

Numbers	Words & Expressions	Meanings	Notes
1	script /skrɪpt/	n. 笔迹；剧本	
2	flourish /ˈflʌrɪʃ/	n. 修饰　v. 繁荣	
3	inscription /ɪnˈskrɪpʃn/	n. 铭文，刻写的文字	
4	rhythm /ˈrɪðəm/	n. 节奏，韵律	
5	stroke /strəʊk/	n. 笔画，一笔	
6	posture /ˈpɒstʃə(r)/	n. 姿态，姿势	
7	grand /grɑːnd/	adj. 宏大的，气派的	
8	changeable /ˈtʃeɪndʒəbl/	adj. 变化的，变幻无常的	
9	calligrapher /kəˈlɪgrəfə(r)/	n. 书法家	
10	tremendous /trəˈmendəs/	adj. 巨大的；极大的	
11	artistic /ɑːˈtɪstɪk/	adj. 艺术的；艺术家的	
12	treasury /ˈtreʒəri/	n. 宝库	
13	dauntless /ˈdɔːntləs/	adj. 无所畏惧的；勇敢的	
14	promising /ˈprɒmɪsɪŋ/	adj. 有希望的；有前途的	
15	large-seal script	大篆	
16	small-seal script	小篆	
17	running script	行书	
18	official script	隶书	
19	regular script	楷书	
20	grass script	草书	

2 Read and learn the key terms of Chinese characters and calligraphy, and tick the ones you know in the last column of the word list.

Numbers	Terms	Meanings	Notes
1	the horizontal and vertical strokes	水平和垂直笔画	
2	culture treasury	文化宝库	
3	the information era	信息时代	
4	be confronted with	面对	
5	dauntless efforts	不屈不挠的努力	
6	handwriting input and pronunciation input	手写输入和语音输入	
7	to gain a second youth	再次繁荣	

3 Get to know some useful expressions for Chinese calligraphy.

(1) This calligraphy is vigorous and forceful.

这部书法作品苍劲有力。

(2) You can really get a lot from a piece of Chinese calligraphy.

从一幅中国书法作品中，你能读出好多东西。

(3) Handwriting is judged by the quality of the brushwork and the abstract beauty of strokes.

鉴赏书法既要看笔底功力又要看笔画的抽象美。

(4) Calligraphy is the most talked-about icon of Chinese culture.

书法是最经典的中国符号。

(5) Calligraphy manifests the basic characteristics of all Chinese arts.

书法比较集中地体现了中国艺术的基本特征。

Task 3 Consolidated Exercises

1 Fill in the blanks according to the passage.

(1) Chinese calligraphy is a(n) _____ oriental art form with a brilliant tradition as ancient as the culture itself.

(2) The present style of Chinese calligraphy is to shape rhythm image with _____.

(3) The lines of regular script can make people feel grand and _____.

(4) The art of Chinese calligraphy with such unique _____ style is also a brilliant peal in the culture treasury of the whole world.

(5) With the further _____ of such technologies as handwriting input and pronunciation input, the ancient Chinese characters have gained a second youth in today's information society and have a more promising future.

2 Translate the following paragraph into English.

书法是一门艺术，可以追溯到中国历史的最早期。虽然书法沟通的工具是汉字，但人们不需要了解中国的语言就能欣赏它的魅力。因为从本质上讲，书法是一门抽象艺术。几千年来，毛笔一直被用于练习和提高人们的手写书法。这已经成为一种中国的传统工艺美术。

3 Please make a brief introduction of Chinese calligraphy by mind map according to the passage.

Chinese
Calligraphy

Section Ⅲ
Let's Practice

Task 1 Let's Talk

1 Please work in pairs to recommend a calligraphy work for decoration of the meeting room in your company, and list the reasons for your recommendation.

_____

_____

_____

_____

2 Read and repeat the dialogue.

A: Hey Allen, I heard that you are really good at Chinese calligraphy, you are so brilliant! How did you do that?

B: Well, my dad is a big fan of Chinese calligraphy. So since I was five, he started to teach me how to draw every single stroke for each character in the correct sequence. I learned new characters every day during the course of the next 15 years.

A: Wow, hard work pays off! You know what, I want to… to be your student. Seriously, can you teach me about Chinese calligraphy? I dream to be just as a good calligrapher as you are one day.

B: Well, since we only have five minutes now, how about meeting up tomorrow afternoon at 3: 00 o'clock. I can share you stories of my favourite calligraphers and their works.

A: Deal!

B: Deal!

A: Hey Allen, you know what, maybe you can tell me one or two names, and I can check online myself first before our meeting tomorrow.

B: Well, I may say Wang Xizhi and Zhang Xu, they are both famous for grass script. Their works are very changeable and full of the abstract beauty. You certainly will like them.

A: Great! Thank you so much man! See you tomorrow!

B: OK, bye!

A: Bye!

1 Imitate the above dialogue first, and then make a conversation with your partner about your favourite calligrapher and his/her works in class.

2 Critical thinking.

Choose one of the following topics, work in groups to translate it into English, then discuss and present your answer in class.

(1) How do you think of the saying "字如其人"?

(2) How do you understand the saying "王羲之写字——入木三分"?

Section Ⅳ
Extended Reading

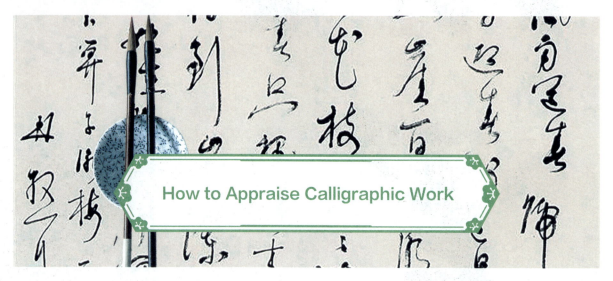

How to Appraise Calligraphic Work

It is very difficult to appraise a calligraphic work. This is because the requirements of a work include the practical function of communicating information through seemingly simple points and lines, the expressive function of communicating feelings, and the aesthetic function. The simpler the expression, the richer the content. The more abstract it is, the deeper its implication. It is a medium for expressing the mood, will, feelings, ideas and the pursuit of beauty of its producer. This requires the appraisers and connoisseurs to have extensive knowledge and keen observation.

Requirements for Calligraphers

True Chinese calligraphers emphasize that the hand should follow the mind when writing. They claim that calligraphers need to concentrate their thoughts and vital energies on the tip of the brush to obtain the perfect state of mind for writing. Only then will their hands be free to wield the brush as the "snake crawls and dragon flies".

A Preface to the Orchid Pavilion Poems

A Preface to the Orchid Pavilion Poems is the best-known example of Wang Xizhi's running hand. It is regarded as the "First Running Hand Under the Sun". It is said that on the third day of the third month of the lunar calendar, in the year 353 (the ninth year of the reign of Emperor Mu of the Eastern Jin Dynasty), Wang Xizhi and a group of eminent scholars gathered at the Orchid Pavilion in Shanyin, Kuaiji, where, according to the conventions of an old sacrificial ceremony, they drank and enjoyed themselves on the bank of a river. When the wine cup flowing on the water stopped before one of them, that person would compose a poem immediately, otherwise he would have to take a drink. Wang wrote a preface to the collection of all these poems: *A Preface to the Orchid Pavilion Poems*. (309 words)

Task 1　New Words and Expressions

Read the following words and expressions, and tick the ones you know in the last column of the word list.

Numbers	Words & Expressions	Meanings	Notes
1	appraise /əˈpreɪz /	*vt.* 估量；估价	
2	aesthetic /iːsˈθetɪk/	*adj.* 审美的, 美学的	
3	implication / ˌɪmplɪˈkeɪʃn /	*n.* 暗指, 含意	
4	connoisseur / ˌkɒnəˈsɜː(r) /	*n.* 鉴赏家；鉴定家；行家	
5	wield /wiːld /	*vt.* 挥、使用（武器、工具等）	
6	pavilion /pəˈvɪliən/	*n.* （公园中的）亭, 阁	
7	eminent /ˈemɪnənt /	*adj.* 非凡的, 杰出的	
8	sacrificial / ˌsækrɪˈfɪʃl /	*adj.* 用于祭献的	

Task 2　Comprehension

After reading the passage, please make a brief introduction of the famous calligraphy work *A Preface to the Orchid Pavilion Poems* in your own words.

_____

_____

_____

_____

Insight into Proverbs

"Li" 礼—Etiquette

China has been known as the "home of etiquette" for thousands of years. Ritual is the law of the person to advance and retreat in order to get what he is. Respecting the old and loving the young, having rules to follow, not committing adultery, and not defeating human relations are the principles of a gentleman. Always be just, polite, respectful and courteous.

1. 一日为师，终身为父。

Whoever is your teacher, even for a day, consider your father (to respect and care for) your whole life.

2. 君子之修身，内正其心，外正其容。

A gentleman's self-cultivation is to uphold his heart and face.

3. 德行广大而守以恭者荣。

A man of great virtue is honored by his deference.

4. 不学礼，无以立。

No etiquette, no standing.

5. 礼貌不须什么代价。

Civility costs nothing.

6. 若要好，问三老。

If you want (to do things) well, ask the old (for advice).

7. 静以修身，俭以养德。

Quiet to cultivate oneself, thrifty to cultivate morality.

8. 千里送鹅毛，物轻人意重。

Giving goose feathers from a thousand miles is a matter of great importance.

9. 长于礼仪，而陋于知人心。

Good at etiquette, but bad at knowing people.

10. 礼义生于富足，盗窃起于贫穷。

Righteousness is born of wealth, and theft of poverty.

11. 贵族重权利，百姓重礼仪。

Nobles value rights and people value etiquette.

12. 礼貌周全不花钱，却比什么都值钱。

Politeness costs nothing, but it's worth more than anything.

13. 不敬他人，是自不敬也。

To disrespect others is to disrespect oneself.

14. 入乡随俗。

When you enter a country, follow its customs.

(When in Rome, do as the Romans do.)

15. 躬亲示范。

Teach others by your example.

Section VI
Self-assessment Checklist

1 Now, it's time for you to review your performance after learning this unit. Carry out a self-assessment by checking the following table.

Items	Ratings			
1. Knowledge	**A**	**B**	**C**	**D**
I know the origin and development of Chinese calligraphy.				
I know the roles of calligraphy in Chinese people's life.				
I master the methods of how to appraise Chinese calligraphy properly.				
I master the expressions of Chinese calligraphy.				
2. Skills	**A**	**B**	**C**	**D**
I can use logic to grasp the main idea and detail of the passage.				
I can use the reading skills to grasp the main idea and details of the passage.				
I can guess the meanings of the new words by context.				
I know the basic communication skills and strategies in talking about classic Chinese calligraphers and their works.				
3. Speaking	**A**	**B**	**C**	**D**
I can talk about classic Chinese calligraphers and their works with westerners fluently.				
I can explain the ways of appraising Chinese calligraphy works in English properly.				
I can tell the components which make a Chinese calligraphy work tasteful and worthwhile to appreciate in English.				
I can explain the aesthetic beauty calligraphers showed in Chinese paintings in English.				
4. Confidence in Chinese Culture	**A**	**B**	**C**	**D**
I can understand the connotation in Chinese calligraphy.				
I have the awareness of writing good handwriting both in Chinese and English.				
I feel proud of the culture of Chinese calligraphy.				

A: Basically agree

B: Agree

C: Strongly agree

D: Disagree

2 Fill in the blanks in the mind map below to check whether you have a good understanding of this unit.

(1) Mind map of the Chinese calligraphy.

small-seal script

(2) Think and discuss. Recommend a calligraphy work for a western client (male, 45 years old) who has great interest in music.

_____

_____

_____

_____

_____

_____

_____

_____

_____

_____

_____

_____

Unit 4

Chinese Painting and Poems

What do you know about Chinese painting? How many Chinese poems can you recite? As one of the oldest artistic traditions in the world, Chinese painting in the traditional style is known today in Chinese as guó huà (国画). Chinese painting is the art of brush and ink. The basic tools are almost the same with those of calligraphy, which influenced painting in both technique and theory. Chinese poem is one of the major parts in Chinese culture and it is closely related to Chinese painting. What the poems describe can be drawn in a Chinese painting. For more knowledge about Chinese painting and poems, let's scan the overview to know the details.

Overview

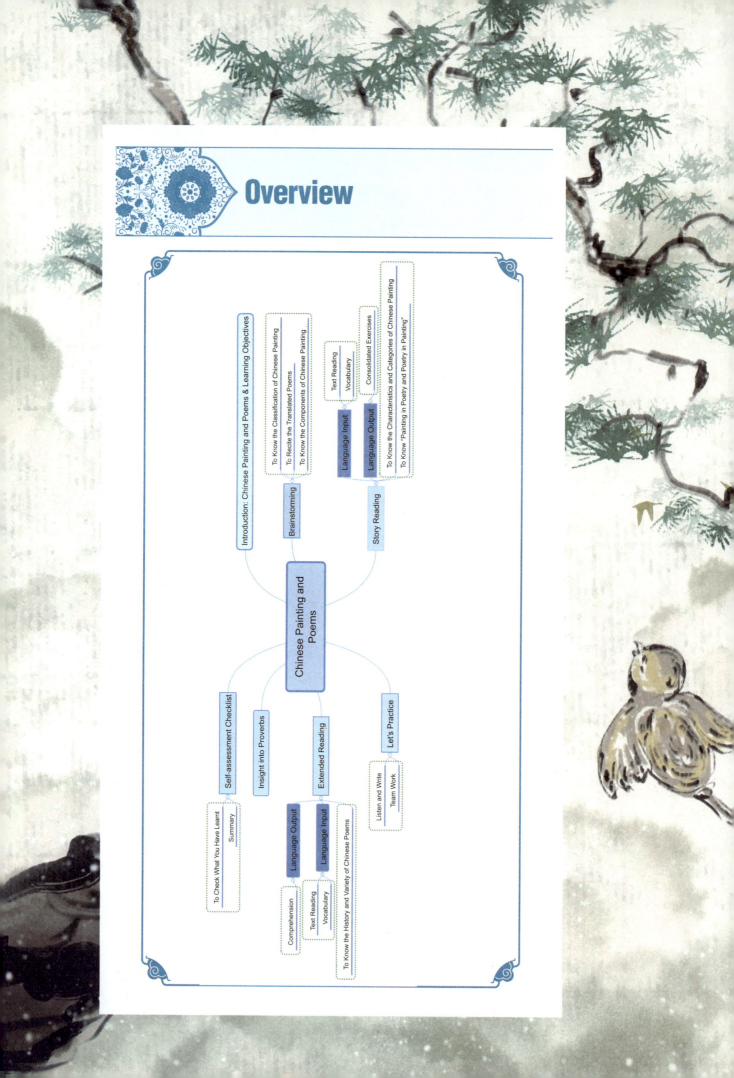

Chinese Painting and Poems

Introduction: Chinese Painting and Poems & Learning Objectives

Brainstorming
- To Know the Classification of Chinese Painting
- To Recite the Translated Poems
- To Know the Components of Chinese Painting

Story Reading
- Language Input
  - Text Reading
  - Vocabulary
  - Consolidated Exercises
- Language Output
  - To Know the Characteristics and Categories of Chinese Painting
  - To Know "Painting in Poetry and Poetry in Painting"

Self-assessment Checklist
- To Check What You Have Learnt
- Summary

Insight into Proverbs

Extended Reading
- Language Output
  - Comprehension
- Language Input
  - Text Reading
  - Vocabulary
  - To Know the History and Variety of Chinese Poems

Let's Practice
- Listen and Write
- Team Work

Learning Objectives

After learning this unit, students are able to reach the goals below.

1	专业能力目标	① 了解中国诗词曲赋和中国画的起源、分类、发展历程及其在人们生活中的影响和作用
		② 通过阅读文章、赏析诗歌掌握与中国诗词曲赋和中国画相关的词汇与句型表达；提升对英文诗歌的创作与赏析能力
		③ 能够提升阅读速度，快速掌握文章的主旨和大意；能理解中国诗歌的发展历史；能分析文章的思想观点、篇章布局、语言技巧及文体修辞等；提高阅读理解能力和表达能力
2	方法能力目标	① 运用诗歌赏析的技巧阅读中国古代、现代诗歌，理解其主旨和大意，领悟各诗节传达的思想情感，掌握全诗体现的思想观点
		② 养成自主阅读的好习惯；通过中国诗词曲赋和中国画的专题阅读摄取诗歌、国画中蕴含的信息；总结要点并制作专题阅读报告；尝试绘画与写诗
		③ 在限定的时间内，运用略读、寻读、推测等方法获取诗词曲赋画中的关键信息，了解事实与观点
		④ 学会在阅读中通过构词法、上下文联系等方法猜测单词意思，扩大词汇量
3	社会能力目标	① 通过广泛的阅读，增强发散性思维与批判性思维；提高想象力、创新能力、社会适应能力
		② 在信息技术时代，提升获取信息和分析信息的能力
		③ 掌握英语快速阅读的技巧，提高学习效率和社会工作效率
		④ 通过阅读扩大知识面；恰当地运用基本知识和语言讲述中国经典诗词曲赋和中国画里的国人智慧；运用中国经典诗词中蕴含的丰富哲理分析解决现实问题
4	情感与思政目标	① 通过本单元的学习，对中国经典诗词曲赋和中国画有更深入的了解；在跨文化交际中能正确地运用中国的经典诗词；激发诗意和浪漫的情怀；感悟中华民族优秀的传统文化
		② 通过主题阅读，熟练运用中国经典诗词、名句、名段、名典；陶冶性情，提高文化涵养和生活品质；提升"四个自信"
		③ 通过主题阅读及拓展阅读，懂得以诗画表达对美好生活的向往、抒发热爱祖国的情怀；增强不同文化的感悟力；提升文化的自觉性

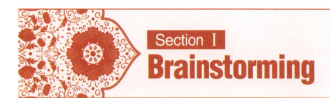

Section I
Brainstorming

Task 1   Let's Warm Up

1 Match the pictures with the correct choice provided.

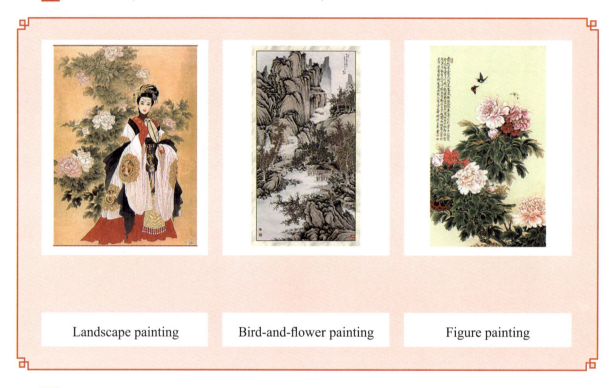

| Landscape painting | Bird-and-flower painting | Figure painting |

2 Recite the Chinese poems according to their English translations.

(1) Friendships across the world make near neighbours of far horizons.

(2) If you would ask me how my sorrow has increased, just see the over-brimming river flowing east!

(3) How rare the moon, so round and clear!

With cup in hand, I ask of the blue sky,

"I do not know in the celestial sphere

What name this festive night goes by?"

(4) Over old trees wreathed with rotten vines fly evening crows;

Under a small bridge near a cottage a stream flows;

On ancient road in the west wind a lean horse goes.

Westward declines the sun;

Far, far from home is the heartbroken one.

Task 2 **Let's Watch and Say**

1 Watch the video about Chinese painting and try to write down the missing words in the blanks.

Red, green, black, white, ____(1)____, light, dry and wet changes of water and ink on paper show the mysteries of traditional Chinese painting—*Guohua*. Tools and materials involve writing ____(2)____, ink, ____(3)____, rice paper and ____(4)____. The hardness or ____(5)____ of brushes, paper absorbency and color, determine the features of *Guohua*.

2 Critical thinking.

(1) Do you know why there always appear high mountains and flowing water in landscape paintings?

(2) As we know, there are many images which embody poet's feelings in Chinese poems. Can you list some and illustrate its embodiment?

_____

_____

_____

_____

_____

_____

_____

Section II
Story Reading

Task 1   Let's Read

Chinese Painting

Looking at Chinese paintings for the first time, you are easy to find it very different from western oil paintings which are devoid of brushwork. Traditional Chinese painting is a combination in the same picture of the arts of poetry, calligraphy, painting, and seal engraving. In ancient times most artists were poets and calligraphers. Su Dongpo (1037—1101) is an example. To the Chinese, "painting in poetry and poetry in painting" has been one of the criteria for excellent works of art. Inscriptions and seal impressions help to explain the painter's ideas and sentiments and also add decorative beauty to the painting. Ancient artists liked to paint pines, bamboos, and plum blossoms. When inscriptions like "exemplary conduct and nobility of character" were made, those plants were meant to embody the qualities of people who were upright and were ready to help each other under hard conditions. For Chinese graphic art, poetry, calligraphy, painting, and seal engraving are necessary parts, which supplement and enrich one another.

According to the means of expression, Chinese painting can be divided into two categories: the Xieyi school and the Gongbi school. The Xieyi school is marked by exaggerated forms and freehand brush work. The Gongbi school is characterized by close attention to detail and fine brush work.

Autumn Evening in My Mountain Abode

Blank hills look pure as a recent rain refines,

As dusk is falling autumn is felt in the bones.

A silvery moon is shining through the pines.

The limpid brooks are gurgling o'er the stones.

Bamboos laugh out as girls from washing whirl.

The lotus stirs where boats out fishing curl.

The scents of spring may go; that's Nature's will.

This season here attracts the noble still.

(288 words)

Task 2　Let's Learn

1　Read the following words and expressions, and tick the ones you know in the last column of the word list.

Numbers	Words & Expressions	Meanings	Notes
1	devoid /dɪˈvɔɪd/	*adj.* 完全没有的；缺乏的	
2	brushwork /ˈbrʌʃwɜːk/	*n.* (画家的)笔触，画法	
3	calligrapher /kəˈlɪgrəfə(r)/	*n.* 书法家	
4	criteria /kraɪˈtɪəriə/	*n.* 标准，准则，原则	
5	inscription /ɪnˈskrɪpʃn/	*n.* 题词；刻写的文字；碑文	
6	sentiment /ˈsentɪmənt/	*n.* 观点；看法；情绪	
7	decorative /ˈdekərətɪv/	*adj.* 装饰性的；做装饰用的	
8	ancient /ˈeɪnʃənt/	*adj.* 古代的；古老的	
9	pine /paɪn/	*n.* 松树　*vt.* 难过，悲伤	
10	exemplary /ɪgˈzempləri/	*adj.* 典范的；可作楷模的	
11	nobility /nəʊˈbɪləti/	*n.* 高贵品质	
12	embody /ɪmˈbɒdi/	*vt.* 体现，代表；包含	
13	upright /ˈʌpraɪt/	*adj.* 正直的；诚实的	
14	graphic /ˈgræfɪk/	*n.* 图案，图形　*adj.* 绘画的；书法的	
15	supplement /ˈsʌplɪmənt/	*n.* 补充物；添加剂　*v.* 增补，补充	
16	exaggerated /ɪgˈzædʒəreɪtɪd/	*adj.* 夸张的；夸大的	
17	freehand /ˈfriːhænd/	*adj.* 徒手画的	
18	abode /əˈbəʊd/	*n.* 住所；家	
19	refine /rɪˈfaɪn/	*v.* 精炼；提取；去除杂质	

Numbers	Words & Expressions	Meanings	Notes
20	limpid /ˈlɪmpɪd/	*adj.* 清澈的；透明的	
21	brook /brʊk/	*n.* 小溪；小川；小河	
22	gurgle /ˈgɜːgl/	*v.* 发潺潺流水声 *n.* 潺潺声	
23	whirl /wɜːl/	*v./n.* 旋转；回旋；急转	
24	stir /stɜː(r)/	*v./n.* 搅拌；搅和	
25	curl /kɜːl/	*v.* 卷；拳曲 *n.* 鬈发；卷曲物	
26	scent /sent/	*n.* 香味；气息；气味	
27	be divided into	分为；划分	
28	be marked by	以……为特征；具有……特点	
29	be characterized by	以……为特征；特点是……	

2 Read and learn the key terms of Chinese painting, and tick the ones you know in the last column of the word list.

Numbers	Terms	Meanings	Notes
1	seal engraving	刻印	
2	seal impression	印鉴	
3	plum blossoms	梅花	
4	the Xieyi school	写意派	
5	the Gongbi school	工笔派	
6	painting in poetry and poetry in painting	诗中有画, 画中有诗	

3 Get to know some useful expressions for etiquette of Chinese paintings.

According to the traditional culture—"fengshui", there are some taboos in drawing a Chinese painting.

(1) Plum blossom cannot be drawn in a reverse direction because the reverse plum blossom means "倒梅" in Chinese, homophonic with "倒霉" (bad luck in English).

(2) Black crow often doesn't appear in a Chinese painting because in Chinese culture black crow is a bird of ill omen symbolizing misfortune or doom.

(3) Withered poplar and willow (枯杨败柳) usually cannot be a single image in a Chinese painting, because it represents decline or downfall.

Task 3 Consolidated Exercises

1 Answer the following questions briefly according to the passage.

(1) What are the components of a Chinese painting?

(2) What is the criteria for excellent works of art in Chinese painting?

(3) What plant did ancient artists love to paint in Chinese painting? Can you list more?

(4) How is Chinese painting divided? What are the characteristics of those two schools?

(5) Recite the translated poem in Chinese. Who wrote it? What do you know about the poet?

(6) Mark the images in the painting which was illustrated in the poem.

2 Fill in the blanks with the proper words given, and change the word form if necessary.

criteria	embody	upright	supplement
refine	stir	be divided into	be characterized by

(1) Their daily work is to _____ crude oil.

(2) Leadership can _____ one's position in an organization.

(3) Governments that _____ human rights must champion them in their foreign policies—in all places, for all peoples and at all times.

(4) No candidate fulfils all the _____ for this position.

(5) As with any vitamin or _____, please consult your doctor before taking.

(6) A quick _____ will mix them thoroughly.

(7) Colors can _____ two general categories: Cool and Warm.

(8) He was sitting on a(n) _____ chair beside his bed, reading.

3 Translate the following sentences into English using patterns given in the brackets.

(1) 他在他的画中体现了新思想。(embody)

(2) 糖、油和金属在使用前须先提炼。(refine)

(3) 把汤搅拌一下。(stir)

(4) 这里的标准将不同于其他地方适用的标准。(criteria)

(5) 请把您的座位靠背转到垂直位置。(upright)

(6) 她给私人授课以增加收入。(supplement)

(7) 所有的调查员将被分为三个小组。(be divided into)

(8) 智能手机市场的快速增长是中国手机市场的突出特征。(be characterized by)

Section III
Let's Practice

Task 1 Let's Talk

1 Listen to the story and write down the proper answers in the blanks.

A Brief Introduction to Chinese Painting

Chinese painting has a long history of more than 2,000 years. Three major kinds of subject ____(1)____ dominate Chinese painting. They are birds and flowers, ____(2)____, and landscapes of the countryside, mountains and sea. Chinese painting is also closely related to the art of fine handwriting called calligraphy. Chinese painters use black ink to produce different ____(3)____ and a brush to make many kinds of ____(4)____. Even when they added color, the ink drawing still served as the ____(5)____ of different designs. When judging paintings, the Chinese paid more attention to the brushstrokes than to the subject matter.

Chinese painting, also known as the traditional national painting in China, has its unique and independent ____(6)____ in the world's fine arts field. Using brushes, ink, and different pigments, a painting is ____(7)____ on a special kind of paper. As aforementioned, the traditional subjects in these paintings are figures, landscapes, flowers and birds. They are divided into two different ____(8)____: one is Gongbi, or meticulous painting, the traditional realistic style ____(9)____ by fine brushwork and close attention to detail. The other is Xieyi, or impressionist painting, the ____(10)____ brushwork style characterized by vivid expression and bold and vigorous outlines. The forms of painting include wall paintings, screens, scrolls, albums, and fan covers.

2 Appreciate the poem below, read it with appropriate rhythm and try to draw a picture with the image shown in the poem.

Goodbye Again, Cambridge

(Xu Zhimo)

I leave softly, gently,

Exactly as I came.

I wave to the western sky,

Telling it goodbye softly, gently.

The golden willow at the river edge

Is the setting sun's bride.

Her quivering reflection

Stays fixed in my mind.

Green grass on the bank

Dances on a watery floor

In bright reflection.

I wish myself a bit of waterweed

Vibrating to the ripple. Of the River Cam.

That creek in the shade of the great elms,

Is not a creek but a shattered rainbow, printed on the water

And inlaid with duckweed, it is my lost dream.

Hunting a dream? Wielding a long punting pole

I get my boat into green water, into still greener grass.

In a flood of starlight, on a river of silver and diamond

I sing to my heart's content.

But now, no, I cannot sing

Quietness is my farewell music.

Even Summer insects heap silence for me

Silent is Cambridge tonight!

I leave quietly

As I came quietly.

Gently I flick my sleeves.

Not even a wisp of cloud will I bring away.

Task 2 **Team Work**

Divide students into several groups and ask them to draw a Chinese painting according to one of the poems learnt in middle school.

Section IV
Extended Reading

Chinese Poems

Chinese poetry can be divided into three main periods: the early period, characterized by folk songs in simple, repetitive forms; the classical period from the Han Dynasty to the fall of the Qing Dynasty, in which a number of different forms were developed; and the modern period of Westernized free verse.

Early poetry

The *Shijing* (《诗经》 literally *Classic of Poetry*, also called *Book of Songs*) was the first major collection of Chinese poems, collecting both aristocratic poems (Odes) and more rustic poetry, probably derived from folk songs (Songs).

A second, more lyrical and romantic anthology was the *Chuci* (《楚辞》 *Elegies of Chu*), made up primarily of poems ascribed to the semi-legendary Qu Yuan (340 BC—278 BC) and his follower Song Yu (fourth century BC).

Classical poetry

During the Han Dynasty (206 BC—220 AD), the Chu lyrics evolved into the fu (prose poetry), a poem usually in rhymed verse except for introductory and concluding passages that are in prose, often in the form of questions and answers; often called a poetical essay (i.e. Robert van Gulik). One of the fine examples of the fu is Ji Kang's *Qin Fu* (《琴赋》), or *Poetical Essay in Praise of the Qin*.

From the Han Dynasty onwards, a process similar to the origins of the *Shijing* produced the Yuefu poems. Again, these were song lyrics, including original folk songs, court imitations and versions by known poets (the best known of the latter being those of Li Bai).

From the second century AD, the Yuefu began to develop into shi or classical poetry—the form which was to dominate Chinese poetry until the modern era. These poems have five or seven character lines, with a caesura before the last three characters of each line. They are divided into the original Gushi (a form of pre-Tang poetry) and Jintishi (modern style poetry), a stricter form developed in the Tang Dynasty with rules governing tone patterns and the structure of the content. The greatest writers of Gushi and Jintishi are often held to be Li Bai and Du Fu respectively.

Towards the end of the Tang Dynasty, the ci poetical form became more popular. Most closely associated with the Song Dynasty, ci most often expressed feelings of desire, often in an adopted persona, but the greatest exponents of the form (such as Li Yu and Su Shi) used it to address a wide range of topics.

As the ci gradually became more literary and artificial after Song times, the sanqu, a freer form, based on new popular songs, developed. The use of sanqu songs in drama marked an important step in the development of vernacular literature.

Later classical poetry

After the Song Dynasty, both poems and lyrics continued to be composed until the end of the imperial period, and to a lesser extent to this day. However, for a number of reasons, these works have always been less highly regarded than those of the Tang Dynasty in particular. Firstly, Chinese literary culture remained in awe of its predecessors: in a self-fulfilling prophecy, writers and readers both expected that new works would not bear comparison with the earlier masters. Secondly, the most common response of these later poets to the tradition which they had inherited was to produce work which was ever more refined and allusive; the resulting poems tend to seem precious or just obscure to modern readers. Thirdly, the increase in population, expansion of literacy, wider dissemination of works through printing and more complete archiving vastly increased the volume of work to consider and made it difficult to identify and properly evaluate those good pieces which were produced. Finally, this period saw the rise of vernacular literature, particularly drama and novels, which increasingly became the main means of cultural expression.

Modern poetry

Modern Chinese poems (新诗, vers libre) usually do not follow any prescribed pattern. Poetry was revolutionized after the May Fourth Movement when writers try to use vernacular styles closer to what was being spoken rather than previously prescribed forms. Early twentieth-century poets like Xu Zhimo, Guo Moruo and Wen Yiduo sought to break Chinese poetry from past conventions by adopting Western models; for example, Xu consciously follows the style of the Romantic poets with end-rhymes.

Since 1980s, poets like Ai Qing used more liberal running lines and direct diction, which were vastly popular and widely imitated.

In the contemporary poetic scene, the most important and influential poets are the group known as Misty Poets, who use allusion and hermetic references. (760 words)

Task 1　New Words and Expressions

Read the following words and expressions, and tick the ones you know in the last column of the word list.

Numbers	Words & Expressions	Meanings	Notes
1	literally /ˈlɪtərəli/	adv. 按字面；字面上	
2	collection /kəˈlekʃn/	n. 收集物，收藏品	
3	aristocratic /ˌærɪstəˈkrætɪk /	adj. 贵族的	
4	rustic /ˈrʌstɪk /	adj. 乡村的；淳朴的	
5	folksong /ˈfəʊksɒŋ/	n. 民歌，民谣	
6	lyrical /ˈlɪrɪkl/	adj. 抒情的；抒情诗的	
7	anthology /ænˈθɒlədʒi/	n. 选集；文选	
8	semilegendary /semɪˈledʒəndərɪ/	adj. 半传奇的	
9	follower /ˈfɒləʊə(r)/	n. 追随者	
10	rhymed /ˈraɪmd/	adj. 押韵的	
11	verse /vɜːs/	n. 诗节	
12	onwards /ˈɒnwədz/	adv. 向前；在前面	
13	origin /ˈɒrɪdʒɪn/	n. 起源；出身	
14	court /kɔːt/	n. 法院；庭院；宫廷	
15	imitation /ˌɪmɪˈteɪʃn /	n. 模范；效仿	
16	version /ˈvɜːʃn/	n. 版本；译本；形式	
17	dominate /ˈdɒmɪneɪt /	v. 支配；统治；控制	
18	era /ˈɪərə /	n. 时代；年代	
19	caesura /siˈzjʊərə/	n. 休止；停顿	
20	adopted /əˈdɒptɪd/	adj. 被收养的，被采纳的	
21	persona /pəˈsəʊnə/	n. 人格面具；形象	
22	address /əˈdres/	v. 发表；提出；对付	
23	compose /kəmˈpəʊz/	v. 作曲；创作；编排	
24	imperial /ɪmˈpɪəriəl/	adj.帝国的；皇帝的	
25	awe /ɔː/	n. 敬畏；恐惧　v. 使敬畏	
26	predecessor /ˈpriːdɪsesə(r)/	n. 祖先；前任	
27	prophecy /ˈprɒfəsi/	n. 预言	
28	bear /beə(r)/	v. 忍受，承受	
29	inherit /ɪnˈherɪt/	v. 继承，遗传	
30	allusive /əˈluːsɪv/	adj. 暗示的；引用经典的	
31	obscure /əbˈskjʊə(r)/	v. 使模糊　adj. 难解的	
32	dissemination /dɪˌsemɪˈneɪʃn/	n. 传播；宣传	
33	prescribed /prɪˈskraɪbd/	adj. 规定的	
34	hermetic /hɜːˈmetɪk/	adj. 与外界隔绝的	
35	be derived from	来源于	
36	be ascribed to	归因于	
37	be associated with	与……相联系	
38	vernacular literature	通俗文学	
39	Misty Poets	朦胧诗	

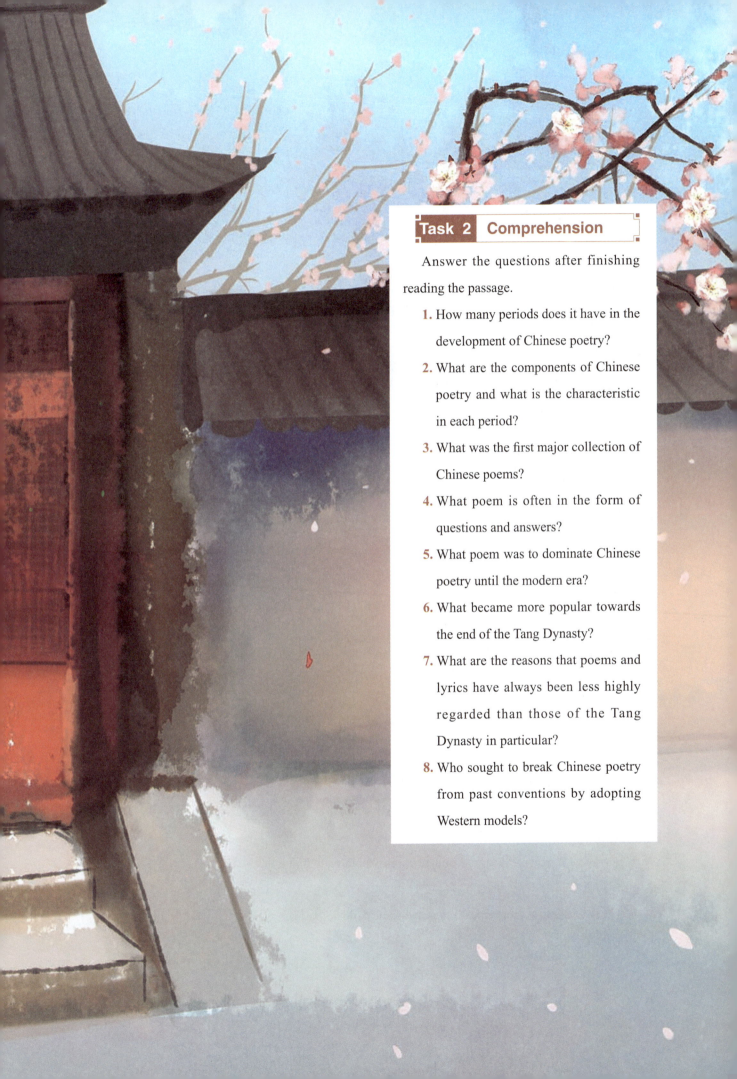

Task 2 Comprehension

Answer the questions after finishing reading the passage.

1. How many periods does it have in the development of Chinese poetry?

2. What are the components of Chinese poetry and what is the characteristic in each period?

3. What was the first major collection of Chinese poems?

4. What poem is often in the form of questions and answers?

5. What poem was to dominate Chinese poetry until the modern era?

6. What became more popular towards the end of the Tang Dynasty?

7. What are the reasons that poems and lyrics have always been less highly regarded than those of the Tang Dynasty in particular?

8. Who sought to break Chinese poetry from past conventions by adopting Western models?

Section V
Insight into Proverbs

"Zhi" 智—Wisdom

A wise man who knows everything is an intelligent person. He knows what is right and wrong, what is good and evil, and what is true and false. He can deal with the problems by Zhongyong (Doctrine of the Mean), wisely.

1. 上善若水，水善利万物而不争。

The highest excellence is like (that of) water. The excellence of water appears in its benefiting all things.

2. 静水流深。

Still water runs deep.

3. 不出户，知天下；不窥牖，见天道。

One can know the world without leaving his door. One can see all the Ways of Heaven without looking out of his window.

4. 智者当差，不用交代。

Send a wise man on an errand, and he will say nothing to him.

5. 知者不言，言者不知。

Those who know do not speak; those who speak do not know.

6. 知足常乐。

He who once knows the contentment that comes simply through being content, will never again be otherwise than contented.

(Do not procrastinate, but use your time wisely from the beginning.)

7. 愚者不问，问者不愚。

The fool does not ask; he who asks is no fool.

(If you really want to learn, you have to be humble enough to

ask questions and reveal your ignorance.)

8. 蠢人嚼舌；智者动脑。

Foolish people wag their tongues; wise people use their brains.

9. 水至清则无鱼。

If water is too clear (and pure), you can't raise fish.

(You can't be too trivial or too inflexible about small matters.

Compromise is often required to get what you want.)

10. 大巧若拙。

Cats hide their paws.

11. 知人者智，自知者明。

To understand others is to have wisdom; to understand oneself is to be illumined.

12. 俗人昭昭，我独昏昏。

The world is full of people that shine; I alone am dark.

13. 智者乐水，仁者乐山。

Men of intellectual character delight in water scenery; men of moral character delight in mountain

scenery.

14. 大智若愚。

The greatest wit seems like clumsiness.

15. 智者千虑必有一失，愚者千虑必有一得。

Even a wise person is sure to be mistaken one time out of a thousand; even a fool is sure to get

something right one time out of a thousand.

(Even the smartest person isn't perfect, while even the dumbest person comes up with a good idea

once in a blue moon.)

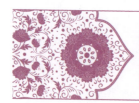

Section Ⅵ
Self-assessment Checklist

1 Now, it's time for you to review your performance after learning this unit. Carry out a self-assessment by checking the following table.

Items	Ratings			
1. Knowledge	A	B	C	D
I know the classification of Chinese painting and its components.				
I know the history and variety of Chinese poems.				
I understand "painting in poetry and poetry in painting".				
I master the expressions of Chinese poems and paintings.				
2. Skills	A	B	C	D
I can use logic to grasp the main idea and details of the passage.				
I can tell the components in a Chinese painting in English.				
I can read a Chinese poem with appropriate rhythm and rhyme.				
I know the skills to draw a Chinese painting and form a Chinese poem.				
3. Speaking	A	B	C	D
I can recite the poems according to their English translations.				
I can recite some proverbs about Zhi.				
I can tell the components in a Chinese painting in English.				
I can explain the images poets show in a Chinese painting.				
4. Confidence in Chinese Culture	A	B	C	D
I can understand the connotation beneath the images shown in a Chinese painting.				
I understand poet's sentiment shown in a Chinese poem.				
I feel proud of the culture of Chinese painting.				

A: Basically agree

B: Agree

C: Strongly agree

D: Disagree

2 Fill in the blanks in the mind maps below to check whether you have a good understanding of this unit.

(1) Mind map of Chinese painting and its classification and components.

(2) Mind map of Chinese poems and its history and variety.

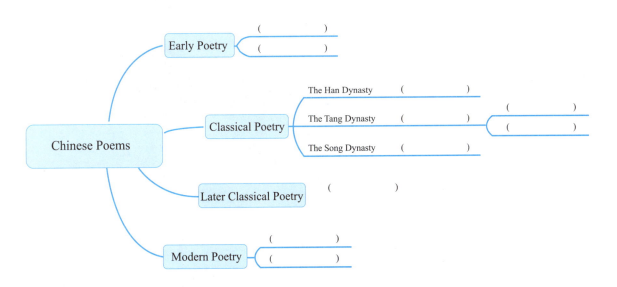

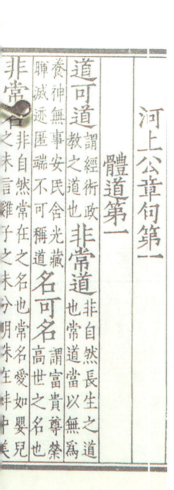

Unit 5

Chinese Classical Literature

Chinese literature has passed down its legacy of magnificent events and works for 3,500 years with a variety of genres and forms like poetry, ci, qu, fu, mythology, novels and drama, reflecting the social climate of its day through the high spirit of art. Generally, Chinese literature is divided into classical literature, modern literature and contemporary literature. Classical literature is represented by the Tang and Song poetry and Four Great Classical Novels.

When we talk about Four Great Classical Novels, something may occur in our mind: Monkey King; Zhuge Liang; Wu Song; Lin Daiyu. We are quite familiar with the characters showing up in the TV series of Four Great Classical Novels. Can you tell what Four Great Classical Novels are? What do you know about Zhuge Liang and his famous writings? In this unit, Chinese classical literature is described. Let's scan the overview to know the details.

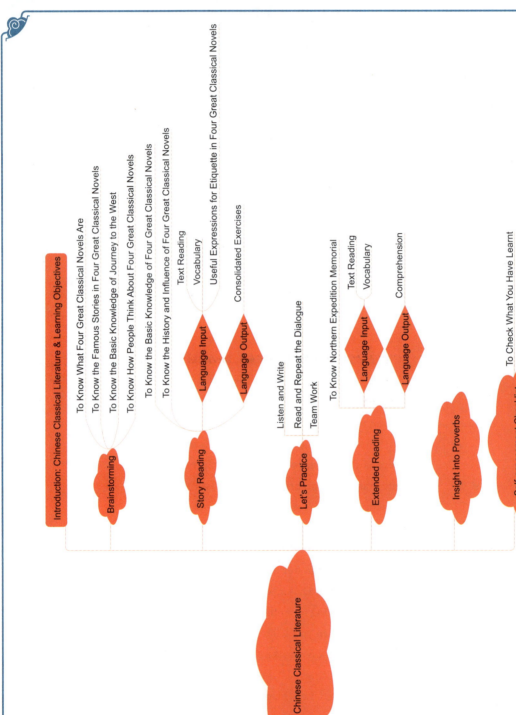

Chinese Classical Literature

Introduction: Chinese Classical Literature & Learning Objectives

Brainstorming
- To Know What Four Great Classical Novels Are
- To Know the Famous Stories in Four Great Classical Novels
- To Know the Basic Knowledge of Journey to the West
- To Know How People Think About Four Great Classical Novels

Story Reading
- To Know the Basic Knowledge of Four Great Classical Novels
- To Know the History and Influence of Four Great Classical Novels
- **Language Input**
  - Text Reading
  - Vocabulary
  - Useful Expressions for Etiquette in Four Great Classical Novels
- **Language Output**
  - Consolidated Exercises

Let's Practice
- Listen and Write
- Read and Repeat the Dialogue
- Team Work

Extended Reading
- To Know Northern Expedition Memorial
- **Language Input**
  - Text Reading
  - Vocabulary
- **Language Output**
  - Comprehension

Insight into Proverbs

Self-assessment Checklist
- To Check What You Have Learnt

Learning Objectives

After learning this unit, students are able to reach the goals below.

1	专业能力目标	① 了解中国经典文学——四大名著的历史、核心内容及其蕴含的人生哲理
		② 学习看图识典故、人物刻画、经典作品赏析等章节,掌握与中国四大名著相关的词汇和句型表达
		③ 通过快速阅读,掌握文章的主旨和段落大意,理解四大名著的核心故事、人物情节,能分析文章的思想观点、篇章布局、语言技巧及文体修辞等,进一步提升阅读理解能力和思想表达能力
2	方法能力目标	① 学会理解文章的主旨和大意,能够分析四大名著的故事情节及人物刻画技巧,掌握阅读材料所体现的哲学思想
		② 养成读图识故事与读故事绘图的习惯,能通过中国四大名著专题阅读,了解故事情节,绘制故事图画,述说故事
		③ 阅读中英文版的《出师表》时,在理解其意义的基础上,通过思维方式的转换,理解中英两种不同语言表达的差异
		④ 在阅读经典文章时,通过构词法、联系上下文语境等方法猜测单词意义,扩大词汇量
3	社会能力目标	① 通过对四大名著基本知识的了解,能在错综复杂的社会环境中,运用批判性思维,看清社会形势,提升社会适应能力
		② 在网络信息时代,依照四大名著诸多经典故事的指引,能辨别真假信息,学会分析信息,做到正确使用信息
		③ 领悟《水浒传》中英雄人物的侠肝义胆,增强社会正义感和弘扬奉献精神
		④ 培养发散性思维,以便能运用基本知识和语言讲述中国四大名著的经典故事和书中呈现的智慧,并能恰当地运用经典故事提升文学素养、丰富精神生活
4	情感与思政目标	① 通过本单元的学习,掌握中国四大名著中的生活哲学;提升对事物的判断能力、思考能力和决策能力
		② 通过主题阅读,增强文化自信;培养正确的爱情观、家庭观、忠义观及对生命的敬畏感;培养爱国爱家的道德情操;提升文化素养
		③ 通过主题阅读及拓展阅读,能从中国传统文学经典中体悟民族文化、传统文化的优越性;能用中国传统文化所蕴含的哲理指导学习、工作、生活;能够提升思辨能力

Section I
Brainstorming

Task 1   Let's Warm Up

1 Let's look at the pictures below and tell the name of the books in which the characters appear.

2 Describe what is illustrated in each picture.

Task 2 Let's Watch and Say

1 Watch the video about *Journey to the West* and try to write down the missing words in the blanks.

Journey to the West is the most famous mythology (神话) in China. It was _____(1)_____ by real historical events. A Buddhist _____(2)_____ named Xuanzang walked through the harsh (严酷的) desert for more than 10,000 kilometers on his pilgrimage (朝圣之旅) to India to obtain Buddhist Sutra (佛经). His first disciple (信徒), named Sun Wukong, is a monkey with great _____(3)_____ powers. He is a righteous (正直的) and brave hero who has _____(4)_____ 72 methods of transformation (变换). Unlike conventional heroes, it departs (违背) from tradition and fights for freedom. His tag is, "Hey, I'm coming!"

He defeated powerful demons (魔鬼) on the pilgrimage and helped Xuanzang _____(5)_____ his objective with Zhu Bajie and Sha Wujing. The pilgrimage represents the discipline of Buddhism. *Journey to the West* tells stories about self-discipline (自律) and overcoming difficulties. It has a profound influence on the Chinese spirit.

2 Critical thinking.

(1) Some people say children should not read the *Water Margin* because the main characters in the book are aggressive who could influence children's behaviours. Do you think so? Why?

(2) If you were a professional hunter, which character of the Four Greatest Classical Novels you would recommend to a foreign trade company of sports protective equipment firm to take the job as a CEO for their trading business development?

_____

_____

_____

_____

_____

Story Reading

Task 1 **Let's Read**

China's Four Great Classical Novels

The Four Great Classical Novels are commonly regarded by scholars as the greatest and most influential pre-modern Chinese fictions. Dating from the Ming and Qing Dynasties, the stories, scenes and characters in them have deeply influenced Chinese people's ideas and values. In chronological order, they are *The Romance of the Three Kingdoms* (14th century), *The Water Margin* (14th century), *Journey to the West* (16th century) and *A Dream of the Red Mansions* (18th century).

Luo Guanzhong, author of *The Romance of the Three Kingdoms*, based his novel on both folk tales and historical records of the conflicts among the kingdoms of Wei, Shu and Wu (220—265). The three kingdoms were established by Cao Cao, Liu Bei and Sun Quan respectively. Written in the early Ming Dynasty, it was the first historical novel in China.

The Water Margin was written by Shi Nai'an. It appeared almost at the same time as *The Romance of the Three Kingdoms*. It is based on folk tales about a band of rebels led by Song Jiang at the close of the Northern Song Dynasty. There are 108 heroes and heroines in the novel, which is a satire on official corruption and savage feudal oppression.

Journey to the West was written by Wu Cheng'en of the Ming Dynasty. It is based on the true story of a monk in Tang Dynasty who made a perilous overland trip to India to fetch back some Buddhist Sutras. In the novel, the monk is guided and protected by Sun Wukong, the "Monkey King", a figure from folk legend. Intriguing of its imaginary stories, *Journey to the West* is a famous Chinese mythological novel.

A Dream of the Red Mansions has 120 chapters, of which the first 80 were written by Cao Xueqin, and the remaining 40 by Gao E. With the tragic love story of Jia Baoyu and Lin Daiyu as the main theme, the novel describes the decline of four feudal noble families. It is a treasure house of information about how aristocratic families lived during the Qing Dynasty. Combining artistry and ideology, it is a great masterpiece of realism in Chinese literature.

The Four Great Classical Novels are among the world's oldest novels and are considered to be the pinnacle of China's achievement in classical novels. They have been influencing the creation of many stories, plays, movies and other forms of entertainment throughout East Asia, in countries such as China, Japan, Republic of Korea, and Vietnam. (412 words)

Task 2　Let's Learn

1　Read the following words and expressions, and tick the ones you know in the last column of the word list.

Numbers	Words & Expressions	Meanings	Notes
1	commonly /ˈkɒmənli/	*adv.* 通常；常常	
2	scholar /ˈskɒlə(r)/	*n.* 学者	
3	influential /ˌɪnfluˈenʃl/	*adj.* 有很大影响的；有支配力的	
4	pre-modern /ˈpriːˈmɒdn/	*adj.* 前现代的	
5	fiction /ˈfɪkʃn/	*n.* 小说；虚构的事	
6	tale /teɪl/	*n.* 故事；历险记	
7	conflict /ˈkɒnflɪkt; kənˈflɪkt/	*n.* 冲突；争执；争论	
8	establish /ɪˈstæblɪʃ/	*v.* 建立；创立；设立	
9	band /bænd/	*n.* 一伙人；一帮人；乐队	
10	rebel /ˈrebl; rɪˈbel/	*n.* 反政府的人；叛乱者；造反者	
11	heroine /ˈherəʊɪn/	*n.* 女英雄	
12	satire /ˈsætaɪə(r)/	*n.* 讽刺；讥讽	
13	corruption /kəˈrʌpʃn/	*n.* 腐败；贪污；贿赂	
14	savage /ˈsævɪdʒ/	*adj.* 凶恶的；凶残的；野蛮的	
15	feudal /ˈfjuːdl/	*adj.* 封建（制度）的	
16	oppression /əˈpreʃn/	*n.* 压迫；压制	
17	perilous /ˈperələs/	*adj.* 危险的；艰险的	
18	overland /ˈəʊvəlænd/	*adj.* 横跨陆地的	
19	Buddhist /ˈbʊdɪst/	*adj.* 佛教的；佛的	
20	sutras /ˈsuːtrəz/	*n.* 经书；信条	
21	intrigue /ɪnˈtriːɡ; ˈɪntriːɡ/	*v.* 激起……兴趣；引发……好奇心	

Numbers	Words & Expressions	Meanings	Notes
22	mythological /ˌmɪθəˈlɒdʒɪkl/	*adj.* 神话的；神话学的	
23	treasure /ˈtreʒə(r) /	*n.* 宝物；珍品	
24	artistry /ˈɑːtɪstri /	*n.* 艺术技巧	
25	ideology /ˌaɪdiˈɒlədʒi/	*n.* 思想；意识形态	
26	masterpiece /ˈmɑːstəpiːs /	*n.* 杰作；名著	
27	pinnacle /ˈpɪnəkl/	*n.* 顶点；鼎盛时期	
28	achievement /əˈtʃiːvmənt/	*n.* 成就；成绩	
29	be regarded as	被……视为；当作	
30	be based on	基于；在……基础上	

2 Read and learn the key terms of Four Great Classical Novels，and tick the ones you know in the last column of the word list.

Numbers	Terms	Meanings	Notes
1	novel	（长篇）小说	
2	the early Ming Dynasty	明初	
3	the Northern Song Dynasty	北宋	
4	Monkey King	美猴王	
5	folk legend	民间传说	

3 Get to know some useful expressions for etiquette of Four Great Classical Novels.

The Four Great Classical Novels are very famous in China and they are influencing the creation of many stories, plays, movies and other forms of entertainment in the world. There are many proverbs in those novels.

(1) Two heads are better than one. / Three cobblers with their wits combined, equal Zhuge Liang the master mind. 三个臭皮匠，赛过诸葛亮。

(2) Sit on top of the mountain to watch the tigers fight. 坐山观虎斗。

(3) Force five passes and slay six captains—win glory in battle. 过五关斩六将。

(4) Ride alone for thousands of miles. 千里走单骑。

(5) Return fully loaded. 满载而归。

(6) Everything is ready except the east wind. (All is ready except what is crucial.) 万事俱备，只欠东风。

(7) Nothing is difficult if you put your heart into it. 世上无难事，只怕有心人。

(8) To be bent on work and exert oneself to the utmost, never stop until one dies. 鞠躬尽瘁，死而后已。

Task 3 **Consolidated Exercises**

1 Answer the following questions briefly according to the passage.

(1) List the Four Great Classical Novels in time order and write the authors down.

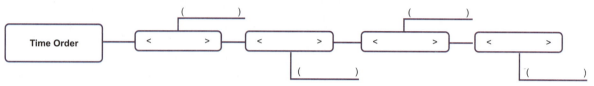

(2) Fill in the brackets with correct kingdoms and their leaders.

(3) What was the creation background when Luo Guanzhong wrote *The Romance of the Three Kingdoms*?

(4) How many heroes and heroines are there in *the Water Margin*? Who led them?

(5) What kind of novel does *Journey to the West* belong to?

(6) What did the novel *A Dream of the Red Mansions* describe?

(7) What kind of novel is *A Dream of the Red Mansions* assigned to?

(8) What is the status of Four Great Classical Novels in the world literature?

2 Fill in the blanks with the proper words given.

(1) The disease is more _____ known as COVID-19. (common)

(2) Earthquake sites should _____ a sad memory rather than a tool to bring in money. (regard as)

(3) Wars and _____ have subjected people living in this area to turbulent life. (conflict)

(4) Our relationship _____ mutual dependence. (base on)

(5) The army should _____ to protect our country. (establish)

(6) We need heroes, and we also need _____. (heroine)

(7) The government has pledged itself to root out _____. (corrupt)

(8) His greatest _____ is his collection of rock records. (treasure)

3 Translate the following sentences into English using patterns given in the brackets.

(1) 这件事情对社会造成了有害的影响。（influence）

(2) 蝙蝠于傍晚时分大批出现。（appear）

(3) 他正努力地忙着写一本新小说。（novel）

(4) 哪里有压迫哪里就有反抗。（oppression）

(5) 丝绸很轻，且易卷，所以可以走陆路运输。（overland）

(6) 本月召开的这次会议，重点集中在文化和意识形态领域。（ideology）

(7) 这真的是一个很值得骄傲的成绩。（achievement）

(8) 人在追求娱乐和消遣之前，必须考虑吃、喝、穿、住。（entertainment）

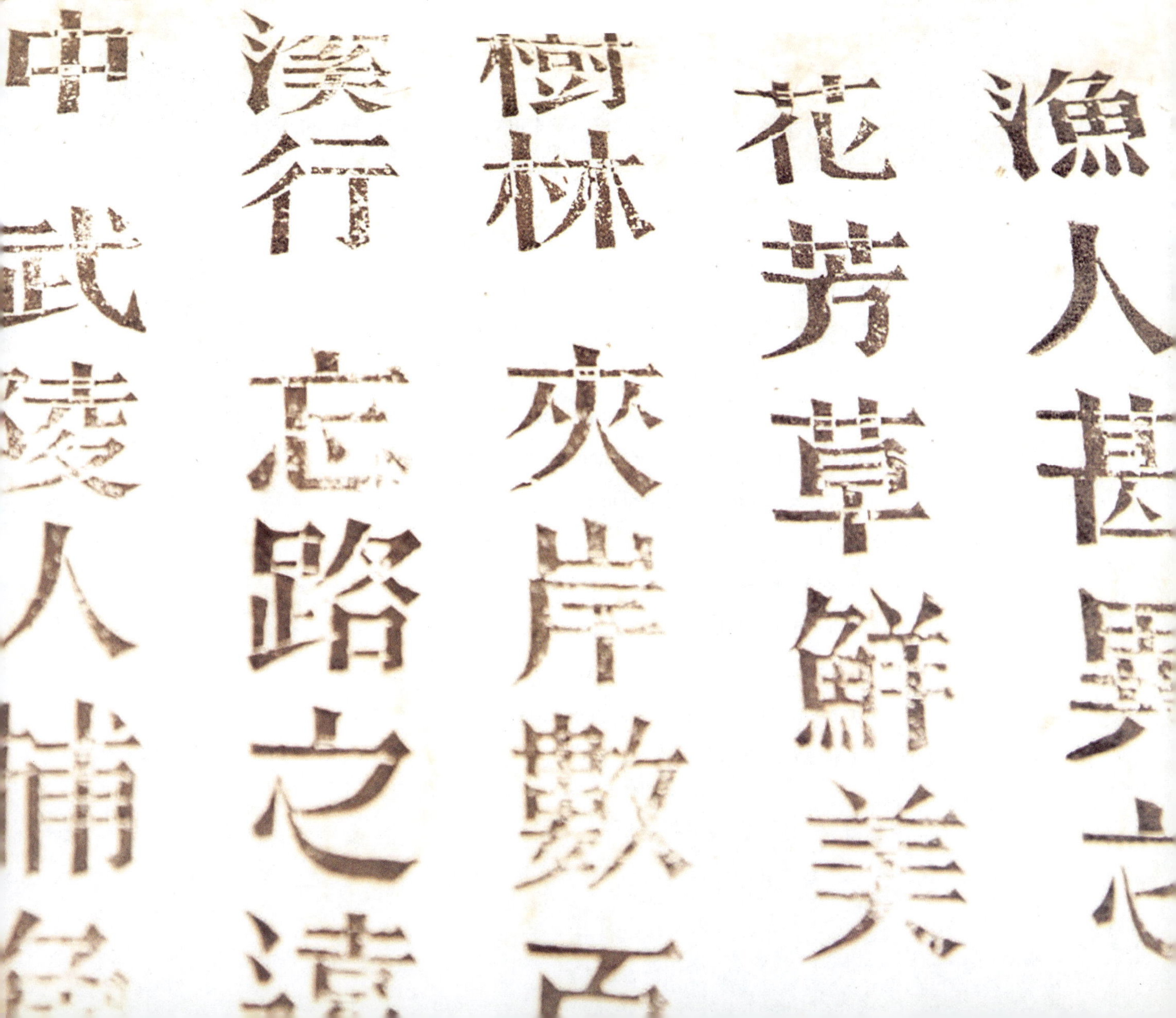

Section III
Let's Practice

Task 1   **Let's Talk**

1 Listen to the story and write down the proper answers in the blanks.

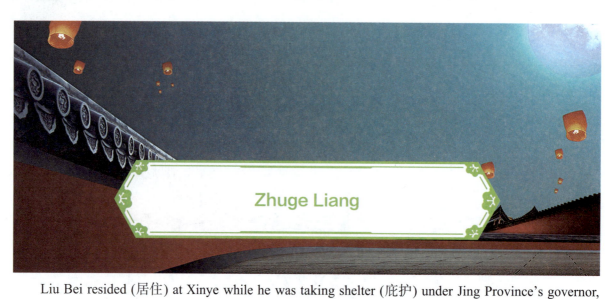

Zhuge Liang

Liu Bei resided (居住) at Xinye while he was taking shelter (庇护) under Jing Province's governor, Liu Biao. Liu Bei visited Sima Hui, who told him, "Confucian academics and ___(1)___ scholars, how much do they know about current affairs? Those who analyze current affairs well are elites (精英). Crouching Dragon and Young Phoenix (卧龙凤雏) are the only ones in this region. " Xu Shu later ___(2)___ Zhuge Liang to Liu Bei again, and Liu wanted to ask Xu to ___(3)___ Zhuge to meet him. However, Xu Shu replied, "You must visit this man in person. He cannot be invited to meet you."

Liu Bei ___(4)___ recruiting (招募) Zhuge Liang after paying three personal visits. Zhuge Liang ___(5)___ the Longzhong Plan to Liu Bei.

Afterwards, Liu Bei became very close to Zhuge Liang and often had ___(6)___ with him. Guan Yu and Zhang Fei were not pleased and ___(7)___. Liu Bei explained, "Now that I've Kongming, it's just like a fish getting into water. I hope you'll stop making unpleasant ___(8)___."Guan Yu and Zhang Fei then stopped complaining.

2 Read and repeat the dialogue.

Wu: Could you imagine? I spent a whole day watching the TV series *Journey to the West* yesterday.

Elizabeth: Oh, really? Is it good?

Wu: It's hugely popular with kids and adults alike. I used to love it when I was little. The Monkey King was my childhood idol!

Elizabeth: What's it about? Is it a fairy story?

Wu: It's an adventure story, based on history. In the Tang Dynasty, Monk Xuanzang was commissioned by the Bodhisattva to travel to the West and fetch back sutras. On the road, he acquired three disciples with great might, and together they fought against evil spirits and demons. Monkey King is his first and mightiest disciple.

Elizabeth: Is that how Buddhism was brought to China?

Wu: No, Buddhism existed in China much earlier, but it became very widespread after Xuanzang.

Task 2 Team Work

1 Role-play.

Imitate the above dialogue first, then make one with your partner talking about Four Great Classical Novels and perform it in class.

2 Group discussion.

(1) Tell the story of the proverb and share your understanding of it: To be bent on work and exert oneself to the utmost, never stop until one dies.

(2) Do you like Monkey King in *Journey to the West*? How do you think of his behavior in *Havoc in Heaven*?

Extended Reading

Northern Expedition Memorial (《出师表》)

Permit me to observe:

The late emperor was taken from us before he could finish his life's work, the restoration of Han. Today, the empire is still divided in three, and our very survival is threatened. Yet still the officials at court and the soldiers throughout the realm remain loyal to you, your majesty. Because they remember the late emperor, all of them, and they wish to repay his kindness in service to you. This is the moment to extend your divine

influence, to honour the memory of the late Emperor and strengthen the morale of your officers. It is not time to listen to bad advice, or close your ears to the suggestions of loyal men.

The court and the administration are as one. Both must be judged by one standard. Those who are loyal and good must get what they deserve, but so must the evil-doers who break the law. This will demonstrate the justice of your rule. There cannot be one law for the court and another for the administration.

Counselors and attendants like Guo Youzhi, Fei Yi, and Dong Yun are all reliable men, loyal of purpose and pure in motive. The late Emperor selected them for office so that they would serve you after his death. These are the men who should be consulted on all palace affairs.

Xiang Chong has proved himself a fine general in battle, and the late Emperor believed in him. That is why the assembly has recommended him for overall command. It will keep the troops happy if he is consulted on all military matters.

The emperors of the Western Han chose their courtiers wisely, and their dynasty flourished. The emperors of the Eastern Han chose poorly, and they doomed the empire to ruin. Whenever the late Emperor discussed this problem with me, he lamented the failings of Emperors Huan and Ling. Advisors like Guo Youzhi, Fei Yi, Chen Zhen, Zhang Yi, and Jiang Wan—these are all men of great integrity and devotion. I encourage you to trust them, your majesty, if the house of Han is to rise again.

I begin as a common man, farming in my fields in Nanyang, doing what I could to survive in an age of chaos. I never had any interest in making a name for myself as a noble. The late Emperor was not ashamed to visit my cottage and seek my advice. Grateful for his regard, I responded to his appeal and threw myself into his service.

Now twenty-one years has passed, the late Emperor always appreciated my caution and, in his final days, entrusted me with his cause. Since that moment, I have been tormented day and night by the fear that I might let him down. That is why I crossed the Lu river at the height of summer, and entered the wastelands beyond. Now the south has been subdued, and our forces are fully armed. I should lead our soldiers to conquer the northern heartland and attempt to remove the hateful traitors, restore the house of Han, and return it to the former capital. This is the way I mean to honor my debt to the late Emperor and fulfill my duty to you. Guo Youzhi, Fei Yi, and Dong Yun are the ones who should be making policy decisions and recommendations.

My only desire is to be permitted to drive out the traitors and restore the Han. If I should let you down, punish my offense and report it to the spirit of the late Emperor. If those three advisors should fail in their duties, then they should be punished for their negligence. Your Majesty, consider your course of action carefully. Seek out good advice, and never forget the late Emperor.

I depart now on a long expedition, and I will be forever grateful if you heed my advice. Blinded by my own tears, I know not what I write. (665 words)

Task 1　New Words and Expressions

Read the following words and expressions, and tick the ones you know in the last column of the word list.

Numbers	Words & Expressions	Meanings	Notes
1	restoration /ˌrestəˈreɪʃn/	n. 恢复；复原；复位	
2	empire /ˈempaɪə(r)/	n. 帝国，大企业	
3	survival /səˈvaɪvl/	n. 生存；存活	
4	threaten /ˈθretn/	v. 危险；恐吓	
5	realm /relm/	n. 王国；领域	
6	majesty /ˈmædʒəsti/	n. 陛下；王权	
7	extend /ɪkˈstend/	v. 扩大；扩展；延长	
8	divine /dɪˈvaɪn/	adj. 天赐的；神圣的	
9	morale /məˈrɑːl/	n. 士气	
10	administration /ədˌmɪnɪˈstreɪʃn/	n. 管理；行政；执行	
11	deserve /dɪˈzɜːv/	v. 值得；应得	
12	evil-doers /ˈiːvl ˈduː(:)əz/	n. 作恶者	
13	demonstrate /ˈdemənstreɪt/	v. 证明；证实；论证	
14	counselor /ˈkaʊns(ə)lə/	n. 律师；顾问；参赞	
15	attendant /əˈtendənt/	n. 侍从；随从	
16	reliable /rɪˈlaɪəbl/	adj. 可靠的；可信赖的	
17	pure /pjʊə(r)/	adj. 纯的；纯净的	
18	motive /ˈməʊtɪv/	n. 动机；原因；目的	
19	select /sɪˈlekt/	v. 选择；挑选	

continued

Numbers	Words & Expressions	Meanings	Notes
20	consult /kən'sʌlt/	v. 咨询，请教	
21	prove /pruːv/	v. 证明；证实	
22	assembly /ə'sembli/	n. 议会；集会；朝会	
23	overall /ˌəʊvər'ɔːl; 'əʊvərɔːl/	adj. 总体的；综合的	
24	command /kə'mɑːnd/	n. 指挥；统率	
25	troop /truːp/	n. 部队；军队	
26	military /'mɪlətri/	adj. 军队的；军事的	
27	doom /duːm/	v. 使……注定失败	
28	lament /lə'ment/	v. 悲痛；惋惜	
29	advisor /əd'vaɪzə/	n. 顾问；提供意见者	
30	integrity /ɪn'tegrəti/	n. 正直；诚信	
31	devotion /dɪ'vəʊʃn/	n. 奉献；忠诚；专心	
32	chaos /'keɪɒs/	n. 混乱；杂乱	
33	ashamed /ə'ʃeɪmd/	adj. 羞愧的；尴尬的	
34	cottage /'kɒtɪdʒ/	n. 小屋；村社	
35	appeal /ə'piːl/	n./v. 呼吁；上诉；申请	
36	caution /'kɔːʃn/	n. 警惕；小心	
37	entrust /ɪn'trʌst/	v. 委托；交托	
38	torment /tɔː'ment/	v. 使折磨；使痛苦	
39	wasteland /'weɪstlænd/	n. 荒地；荒野	
40	subdue /səb'djuː/	v. 征服；压制；克服	
41	traitor /'treɪtə(r)/	n. 背叛者；叛徒	
42	fulfill /fʊl'fɪl/	v. 履行；执行；贯彻	
43	negligence /'neglɪdʒəns/	n. 疏忽；失职	
44	depart /dɪ'pɑːt/	v. 离开；起程；出发	
45	heed /hiːd/	n./v. 注意；留心	

Task 2　Comprehension

Answer the questions after finishing reading the passage.

1. Can you recite one of the paragraphs or the whole passage in Chinese?

2. What did Zhuge Liang think about the court and the administration?

3. Who were the reliable men Zhuge Liang thought?

4. What will keep the troops happy?

5. What determined Han Dynasty to flourish or to ruin?

6. What did Zhuge Liang do before he was recruited?

7. What's the purpose of Zhuge Liang to write *Northern Expedition Memorial*?

8. What was Zhuge Liang's desire?

Section V
Insight into Proverbs

"Xin" 信—Faith

Faith is the antithetical to wisdom: a wise man knows what can be done, while a believer who does not know believes words from others, and then observes manners and practices benevolence. Hence the saying: he who believes is not righteous. In present society, more connotations are paid to faith, like keeping one's promise, integrity, and confidence.

1. 一言既出，驷马难追。

Once a word is spoken, even a team of four horses cannot catch up to it.

(A promise is a promise; the superior person keeps his word and never goes back on it.)

2. 言忠信，行笃敬。

Be conscientious and sincere in what you say; be earnest and serious in what you do.

3. 话经三张嘴，长虫也长腿。

After something has passed through three mouths, even snakes are said to have grown legs.

(Every rumor becomes exaggerated to a ridiculous point and is not to be believed.)

4. 信则人任焉。

If you are trustworthy, men will trust you.

5. 交人交心，浇花浇根。

To make friends, share your true thoughts and feelings, to make flowers grow, you must water their roots.

(To make friends, sincerity is most important.)

6. 人而无信，不知其可也。

I do not know how men get along without good faith.

7. 尽信书不如无书。

Better to go without books than to believe everything they say.

(Don't believe everything you read.)

8. 言为心声。

One's words reflect one's thinking.

9. 一诺千金。

Promise is debt.

10. 言行不一。

Saying is one thing and doing another.

11. 眼见为实。

Seeing is believing.

12. 信言不美，美言不信。

True words are not fine-sounding; fine-sounding words are not true.

13. 知者不失人，亦不失言。

A man of intelligence never loses his opportunity，neither does he waste his words.

14. 人言可畏。

The tongue is not steel, yet it cuts.

15. 巧言乱德。

It is plausible speech which confuses men's ideas of what is moral worth.

Section VI
Self-assessment Checklist

1 Now, it's time for you to review your performance after learning this unit. Carry out a self-assessment by checking the following table.

Items	Ratings			
1. Knowledge	A	B	C	D
I know Chinese classical literature and the four classical ones.				
I know the famous stories from Four Great Classical Novels.				
I know the history and creation background of Four Great Classical Novels.				
I know the English translation of *Northern Expedition Memorial*.				
I master the expressions of Chinese classical literature and the four classical ones.				
2. Skills	A	B	C	D
I can guess the general idea of a paragraph by using topic sentence.				
I can guess the meanings of the new words through word-formation and context.				
I can use synonyms to expand my vocabulary.				
3. Speaking	A	B	C	D
I can tell the names of Four Great Classical Novels in English.				
I can tell the major stories in Four Great Classical Novels.				
I can explain the reasons why Zhuge Liang wrote *Northern Expedition Memorial*.				
I can illustrate Four Great Classical Novels with useful words and expressions in the unit.				
4. Confidence in Chinese Culture	A	B	C	D
I can understand the connotations each story shows in Four Great Classical Novels.				
I understand the philosophies in Four Great Classical Novels.				
I feel proud of Chinese classic works' influence in the world.				

A: Basically agree

B: Agree

C: Strongly agree

D: Disagree

2 Fill in the blanks in the mind maps below to check whether you have a good understanding of this unit.

(1) Mind map of Chinese Four Great Classical Novels.

_____

_____

(2) Mind map of Chinese Four Great Classical Novels and their main characters.

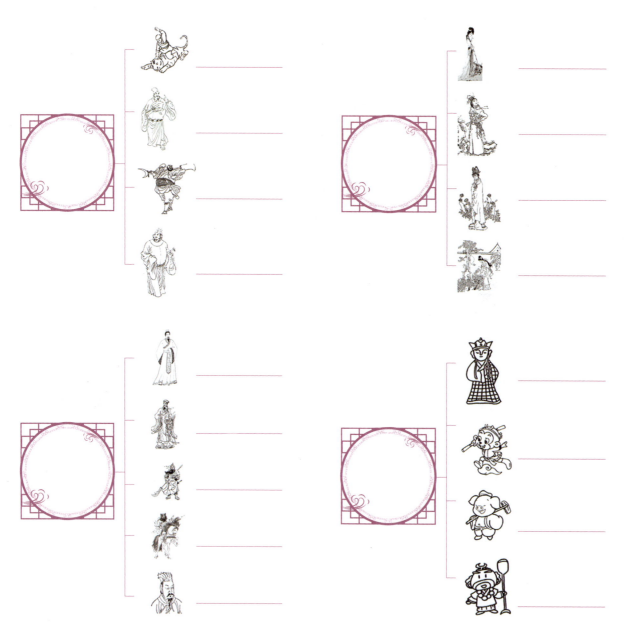

陆

Unit 6

Chinese Opera and Music

Greek tragicomedy, Indian Sanskrit drama, and Chinese opera are indisputably the three most ancient forms of drama in the world. The Chinese word for opera is Xiqu. Chinese opera mainly includes Kun Opera, Qinqiang Opera, Yu Opera, Yue Opera, Huangmei Opera and Peking Opera. They are interesting and attractive for the people from all over the world. Then what makes Chinese opera so special and unique? What roles do they usually have? What techniques do they apply in the play? What musical instruments do they play? This unit will help you open the door of the oriental mysterious culture—Chinese opera. Let's scan the overview to know the details.

Overview

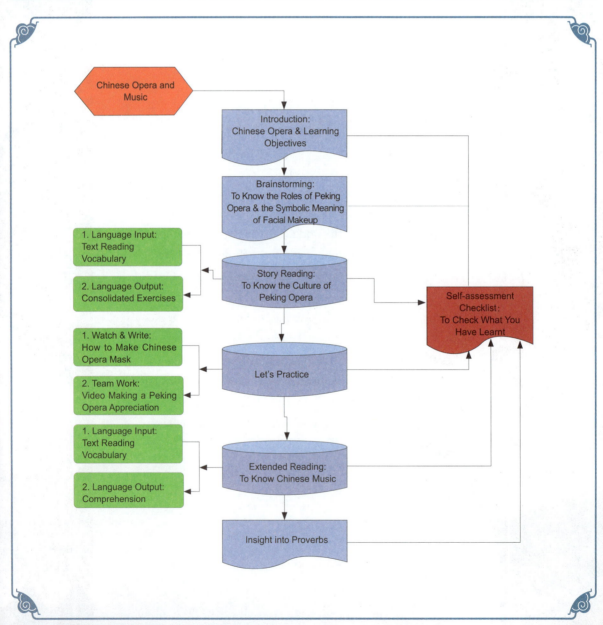

Chinese Opera and Music

Introduction:
Chinese Opera & Learning
Objectives

Brainstorming:
To Know the Roles of Peking
Opera & the Symbolic Meaning
of Facial Makeup

1. Language Input:
Text Reading
Vocabulary

2. Language Output:
Consolidated Exercises

Story Reading:
To Know the Culture of
Peking Opera

Self-assessment
Checklist:
To Check What You
Have Learnt

1. Watch & Write:
How to Make Chinese
Opera Mask

2. Team Work:
Video Making a Peking
Opera Appreciation

Let's Practice

1. Language Input:
Text Reading
Vocabulary

2. Language Output:
Comprehension

Extended Reading:
To Know Chinese Music

Insight into Proverbs

Learning Objectives

After learning this unit, students are able to reach the goals below.

1	专业能力目标	① 了解中国京剧和音乐的起源及其对人们生活的影响
		② 掌握与中国京剧和音乐相关的词汇、句型表达
		③ 提高阅读速度,掌握文章主旨,了解京剧弹唱特点、主题、四大角色的名称、脸谱的特征及含义;能用较流利的语言传播京剧文化
		④ 了解中国传统音乐、乐器的发展;了解古琴在古代中国文人文化生活中的作用和地位
2	方法能力目标	① 运用对比方法提高阅读理解能力
		② 在阅读中能辨别表达事实与观点的句子
		③ 在阅读中通过合成构词法猜测词汇大意,并能触类旁通,扩大词汇量
		④ 养成自主阅读的好习惯,能根据中国戏曲专题阅读摄取相关信息,扩大知识面
3	社会能力目标	① 通过本单元的学习,了解京剧遵从的儒家和道家所提倡的礼仪和规范,能批判性地将这些礼仪和规范运用于社交中,提高社会适应能力
		② 在信息技术时代,能获取其他类型的戏剧、音乐信息,并具备分析该类信息的能力
		③ 能运用京剧中蕴含的儒家和道家思想,提升学习和工作方面的思考能力
		④ 通过主题阅读,扩大知识面;能运用基本知识和恰当的语言讲述中国经典戏曲桥段和国人智慧;能欣赏京剧文化、领悟京剧内涵;能恰当地运用国粹提升生活质量和社会责任感
4	情感与思政目标	① 通过本单元的学习,提升对中国经典戏曲的认识、鉴赏水平和素养;传承家国情怀;提升传播经典戏曲的交际沟通能力
		② 通过主题阅读,了解京剧中所传达的民族精神和革命文化;提升传承非物质文化的意识和社会责任感
		③ 通过本单元的学习,了解中华民族的革命奋斗历程;陶冶革命情怀和情操;能够奋发图强、报效祖国

Section I
Brainstorming

Task 1 **Let's Warm Up**

Look at the pictures below and match them with the correct choices provided.

Zhengdan	Laodan	Huadan	Wudan

Wusheng	Laosheng	Xiaosheng	Clown

Task 2   Let's Watch and Say

1 Watch the video and think about the facial makeup in Peking Opera. Then work in groups to write down the symbolic meaning of different colored faces.

Red face: _____

Black face: _____

White face: _____

Yellow face: _____

Green face: _____

Blue face: _____

2 Read the following passage to check your answers.

Peking Opera Masks or Beijing Opera Faces or Lianpu means the types of facial make-up or face-painting. In Peking Opera, different types of facial paintings express different significance—red facial painting for loyalty, bravery and brotherhood people while some are related to hypocritical man; black for uprightness, resolution, bravery and wisdom while the others mean dark skin and ugly face; white for insidious, treacherous and opinionated figure while it also means elderly people; yellow for homicidal and tyrannical; blue for mighty and cattiness; green face-painting for a forthright and testy temper; purple for the figures who are upright, mighty, sophistication.

3 Critical thinking.

(1) Where do people usually go to enjoy Peking Opera? How to better appreciate the uniqueness of Chinese culture?

(2) Do you think it a good idea to integrate some Chinese Opera elements into pop music to modernize this traditional art form? Why or why not?

Section II
Story Reading

Task 1　Let's Read

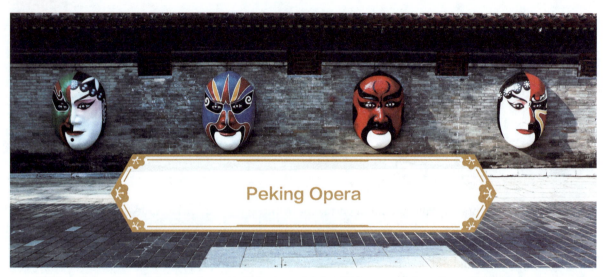

Peking Opera

Peking Opera has a history of nearly 200 years. In the 55th year of the reign of Emperor Qianlong of the Qing Dynasty (1790), four big opera troupes from Anhui Province entered the capital and combined with Kunqu Opera, Yiyang Opera, Hanju Opera and Luantan to create what we now know as Peking Opera. There are currently four main roles in Peking Opera. They are *Sheng*—Male Role, *Dan*—Female Role, *Jing*—Painted Face Male and *Chou*—The Comedy Role.

Peking Opera is a combination of singing, dancing and martial arts. The collaborative costumes and traditional weaponry are combined to tell stories from Chinese folklore, mainly including war and love, everything from warriors fighting in great battles to ordinary people's daily lives. In Qing Dynasty Peking Opera was performed mostly on stage and open-air teahouses and temple courtyards. Since the orchestra played loudly, the performers developed a piercing style of song that can be heard by everyone. The actors use a different singing voice depending on their age. Young female characters usually use piercing falsetto, which needs professional practice.

Mask or face painting is the most unique features of Peking Opera. There are three main features of facial make-ups: contradictory unity of beauty and ugliness, closely related to the role of personality and its stylized design, which helps reveal information about the characters without the need for dialogue. They can represent age, sex, and social standing. Each color has a meaning. Red is heroism and bravery, while white is sinister and mysterious. Facial make-up is different in different situations. *Sheng* and *Dan* have simple facial make-up. They apply little powder, called "handsome play", "plain face", and "clear face". *Jing* and *Chou* have more complex face painting, especially the *Chou* is re-applied paint and design, so called "painted face". Factually, the opera mask is mainly referring to the *Chou* face painting. The *Chou*, because of its dramatic role to play, the nose rubbed a small piece of white powder, commonly known as "small painted face". In 2009, the colored Peking Opera masks were included into the Intangible Cultural Heritage Project list.

Liyuan Theatre, literally meaning pear garden in Chinese and also referring to the place where opera players trained, is the best place to watch Peking Opera. Lao She Teahouse is a Beijing cultural icon. It is named after the celebrated writer of China's famous pre-revolutionary play *Teahouse*. This place's entertained the locals for more than twenty years. Even the former US President George Bush dropped in here to experience Peking Opera for himself.

The best way to enjoy Peking Opera is to close your eyes and listen to the music. You can experience the beauty and moving performance. When you are totally lost in the music, you unconsciously put your hands together and applaud the performer.

As a unique form of Chinese arts, Peking Opera requires the most profound refinement of skills in singing, acting, characterization, movement, choreography and acrobatics. This art form, constituting a valuable part of Chinese culture, has been prevailing all over the world. (540 words)

Task 2 Let's Learn

1 Read the following words and expressions, and tick the ones you know in the last column of the word list.

Numbers	Words & Expressions	Meanings	Notes
1	troupe /truːp/	n. 班团	
2	comedy /ˈkɒmədi/	n. 喜剧	
3	perform /pəˈfɔːm/	v. 表演，执行，完成	
4	orchestra /ˈɔːkɪstrə/	n. 管弦乐队	
5	pierce /pɪəs/	v. (锋利尖锐物体) 刺穿	
6	martial /ˈmɑːʃl/	adj. 军事的；战争的；尚武的	
7	collaborative /kəˈlæbərətɪv/	adj. 合作的，协作的	
8	weaponry /ˈwepənri/	n. 兵器，武器	
9	folklore /ˈfəʊklɔː(r)/	n. 民俗学；民间传说	
10	warrior /ˈwɒriə(r)/	n. 战士，勇士	
11	battle /ˈbætl/	n. 战役；斗争	
12	heroism /ˈherəʊɪzəm/	n. 英勇，英雄气概	
13	sinister /ˈsɪnɪstə(r)/	adj. 阴险的；凶兆的	
14	personality /ˌpɜːsəˈnæləti/	n. 个性；品格	

continued

Numbers	Words & Expressions	Meanings	Notes
15	falsetto /fɔːlˈsetəʊ/	n. 假音；假声歌手	
16	icon /ˈaɪkɒn/	n. 图标；偶像	
17	celebrate /ˈselɪbreɪt/	vt. 庆祝；举行；赞美	
18	revolutionary /revəˈluːʃənəri/	adj. 革命的	
19	unconsciously /ʌnˈkɒnʃəsli/	adv. 不知不觉地；无意识地	
20	applaud /əˈplɔːd/	v. 赞同；称赞；喝彩	
21	require /rɪˈkwaɪə(r)/	vt. 需要；要求；命令	
22	profound /prəˈfaʊnd/	adj. 深厚的；意义深远的	
23	refinement /rɪˈfaɪnmənt/	n. 改进，改善	
24	choreography /ˌkɒriˈɒɡrəfi/	n. 编舞；舞蹈艺术	
25	acrobatic /ˌækrəˈbætɪk/	adj. 杂技的；特技的	
26	to name after	以……命名	
27	to drop in	参加，拜访	
28	to depend on	取决于，依赖	

2 Read and learn the key terms of traditional Chinese opera, and tick the ones you know in the last column of the word list.

Numbers	Terms	Meanings	Notes
1	Peking Opera	京剧	
2	Kunqu Opera	昆剧	
3	Yiyang Opera	弋阳腔，高腔	
4	Hanju Opera	汉剧	
5	Luantan	乱弹	
6	Sheng	生	
7	Dan	旦	
8	Jing	净	
9	Chou	丑	
10	contradictory unity of beauty and ugliness	美与丑的矛盾统一	
11	Mei Lanfang	梅兰芳	
12	Liyuan Theatre	梨园	
13	Lao She Teahouse	老舍茶馆	
14	Intangible Cultural Heritage Project	非物质文化遗产项目	

3 Get to know some useful expressions for etiquette of Peking Opera.

As one of the models of Chinese culture and an important part of Chinese culture, Peking Opera mostly expresses the social life of ancient Chinese, and its art is stylized. It also reflects various kinds of etiquette of ancient society at that time, which can be expressed through specific procedures. Some etiquette that Confucianism emphasized, like the courtesy of the emperor and ministers, colleagues, husband and wife, is seriously displayed on the stage of Peking Opera.

(1) 君臣之礼：行跪拜之礼。

To perform the ritual of kneeling when ministers and other officials went to the court to respect the emperors.

(2) 同僚之礼：行屈膝礼。

To curtsy low when an official meets his colleague whose position is more higher.

(3) 主宾之礼：作揖打躬之礼。

The host and the guests bow to each other with hands folded in front.

(4) 夫妻之礼：作揖打躬之礼。

Couples bow to each other with hands folded in front.

(5) 居家之礼：子女见父母需要躬身施礼，以表对父母的养育之恩。

Children need to bow to their parents to show their gratitude for their upbringing.

(6) 婚丧之礼：通过音乐、服饰、道具的色彩变化，制造出相应的环境气氛。

Wedding and funeral rite: to create the corresponding environmental atmosphere by combining music, costumes and the color props.

A. 娶亲拜堂：红色布置，新娘着红色婚服，吹奏唢呐。

Suona is played to present the wedding meeting the bride in red in the red-decorated hall.

B. 灵堂祭奠：白色布置，祭奠者，身穿孝服，胡琴曲牌伴奏。

Huqin qupai is played to make the mourning atmosphere in the white-decorated hall; the relatives of the death wear Xiaofu.

Task 3   **Consolidated Exercises**

1   Answer the following questions briefly according to the passage.

(1) How many kinds of roles are there in Peking Opera? And what are they?

(2) What are the features of Peking Opera?

(3) What is the main theme of Peking Opera?

(4) What is the singing voice the female performers have to develop and why?

(5) Why is the role of *Chou* also called "painted face"?

(6) How do you know the personality of the performer in Peking Opera?

(7) What honor did Peking Opera masks have in 2009?

(8) Where is the best place to enjoy Peking Opera?

2   Fill in the blanks with the proper words given, and change the word form if necessary.

profound	icon	celebrate	perform
name after	drop in	combine	depend on

(1) Chen Jiaci is a culture _____ representing the Lingnan culture.

(2) I'm looking forward to seeing your _____.

(3) The building is _____ the sponsor.

(4) The origin of this nation is wrapped in _____ mystery.

(5) He _____ his uncle when he was staying in Beijing.

(6) Success _____ not only _____ hardworking but also _____ chances.

(7) The trip will _____ business with pleasure.

(8) He was soon one of the most _____ young painters in England.

3   Translate the following paragraph into English.

梨园源于唐代，它是皇家乐部艺人学艺的场所，表演者被称为"梨园弟子（theatrical performer）"。在清代，京剧逐步形成并在百姓中开始流行。表演是在茶馆、饭馆，甚至是在临时搭建的舞台上进行的。每个演员的脸上画着夸张的图案，代表每个人物的性格、角色和命运。观众可以通过观察人物的脸谱和着装搭配来了解故事。

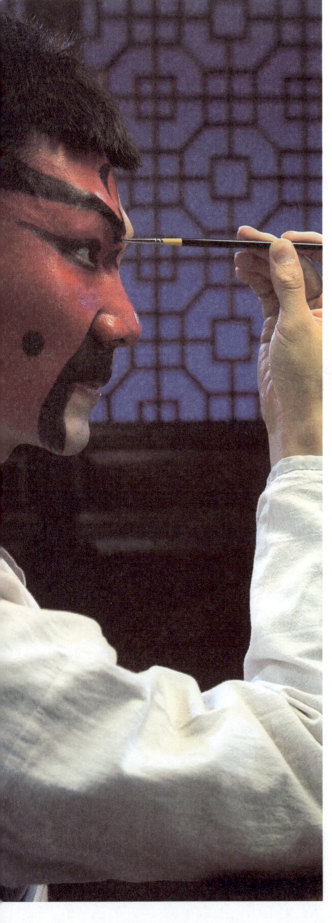

Section Ⅲ
Let's Practice

Task 1   Let's Talk

Watch the video to know the process of making a mask of Peking Opera and try to write them down.

Step 1: _____

Step 2: _____

Step 3: _____

Task 2   Team Work

1   Video making.

Video the process of your team making a mask of Peking Opera. Present it in class to introduce how you make it and what symbolic meaning it conveys.

2   Critical thinking.

Work in groups to discuss the following etiquette in Peking Opera on the stage.

居家之礼：子女见父母需要躬身施礼，以表对父母的养育之恩。(Children need to bow to their parents to show their gratitude for the upbringing.)

3   Peking Opera appreciation.

Each team chooses one piece of the famous Peking Opera works to watch, and then discuss the following questions.

(1) What is the name of the opera?

(2) Who is the key role?

(3) What is the theme of the opera?

(4) What etiquette does the opera embody on the stage?

Section IV
Extended Reading

Traditional Chinese Music

Traditional Chinese music can be traced back to 7,000—8,000 years based on the discovery of a bone flute made in the Neolithic Age. In the Xia, Shang and Zhou Dynasties, only royal families and dignitary officials enjoyed music, which was made on chimes and bells. During the Tang Dynasty, dancing and singing entered the mainstream, spreading from the royal court to the common people. With the introduction of foreign religions such as Buddhism and Islam, exotic and religious melodies were absorbed into Chinese music and were enjoyed by the Chinese people at fairs organized by religious temples.

In the Song Dynasty, original opera such as Zaju and Nanxi was performed in tearooms, theatres, and showplaces. Writers and artists liked it so much that ci (词), a new type of literature resembling lyrics, thrived. During the Yuan Dynasty, qu (曲), another type of literature based on music became popular. This was also a period when many traditional musical instruments were developed such as the pipa, the flute, and the Chinese zither.

Guqin has been the most elegant traditional Chinese musical instrument and the most popular means of self-cultivation by scholars since it was invented by Fu Xi, or by Huang Di over 3,000 years ago. Originally it was played in grand worship ceremonies, later welcomed by all knowledgeable people in Chinese history.

Ancient Chinese people, especially royal families, literati and dignitaries were greatly affected by the doctrines of Confucianism. Confucius (551 BC—479 BC) was a master of guqin. He could sing most ancient poems accompanying guqin, and considered the sound of guqin as the most elegant, appealing music in the world.

Confucius taught that "to educate somebody, you should start with poems, emphasize ceremonies, and finish with music." He thought that a noble man should develop six subjects: *Li* (courtesy), *Yue* (music), *She* (archery), *Yu* (chariot-riding), *Shu* (calligraphy or writing), and *Shu* (computation or mathematics). He ranked music the second place. In Confucian teachings, the purpose and role of music are laid out and the qualities of "good music" are defined. The teachings are seminal for understanding traditional music.

In ancient China, most well-educated people and monks could play classical music as a means of self-cultivation, meditation, mind purification and spiritual elevation, union with nature, identification with the values of past sages, and communication with divine beings or with friends and lovers.

Music was important because the ideal society was to be governed by rites, ritual and ceremonial functions, but not by law or power. In a culture where people function according to ritual and ceremony, music is used to help conduct and govern them. So music wasn't really entertainment, but a means for musicians to accomplish political and social goals. Music was ultimately a means for optimizing social utility or happiness. (478 words)

Task 1　New Words and Expressions

Read the following words and expressions, and tick the ones you know in the last column of the word list.

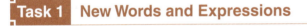

Numbers	Words & Expressions	Meanings	Notes
1	flute /fluːt/	*n.* 长笛	
2	dignitary /ˈdɪgnɪtəri/	*n.* 高官；显要人物	
3	chime /tʃaɪm/	*n.* 钟声；编钟	
4	mainstream /ˈmeɪnstriːm/	*n.* 主流	
5	exotic /ɪgˈzɒtɪk/	*adj.* 异国的；外来的	
6	melody /ˈmelədi/	*n.* 旋律；歌曲	
7	absorb /əbˈzɔːb/	*vt.* 吸收；吸引	
8	literature /ˈlɪtrətʃə(r)/	*n.* 文学；文献	
9	resemble /rɪˈzembl/	*vt.* 类似，像	
10	lyric /ˈlɪrɪk/	*n.* 抒情诗；抒情诗体；歌词	
11	thrive /θraɪv/	*vi.* 繁荣，兴旺	
12	cultivation /ˌkʌltɪˈveɪʃn/	*n.* 培养；耕作	
13	knowledgeable /ˈnɒlɪdʒəbl/	*adj.* 知识渊博的，有知识的	
14	literati /ˌlɪtəˈrɑːti/	*n.* 文人；文学界	
15	affect /əˈfekt/	*vt.* 影响；感染	

Numbers	Words & Expressions	Meanings	Notes
16	elegant /ˈelɪɡənt/	*adj.* 高雅的，优雅的	
17	rank /ræŋk/	*n.* 排；等级	
18	define /dɪˈfaɪn/	*vt.* 定义；使明确	
19	seminal /ˈsemɪnl/	*adj.* 有创造力的；重大的	
20	monk /mʌŋk/	*n.* 僧侣	
21	meditation /ˌmedɪˈteɪʃn/	*n.* 冥想；沉思	
22	purification /ˌpjʊərɪfɪˈkeɪʃn/	*n.* 净化	
23	elevation /ˌelɪˈveɪʃn/	*n.* 高地；海拔	
24	identification /aɪˌdentɪfɪˈkeɪʃn/	*n.* 鉴定，识别	
25	entertainment /ˌentəˈteɪnmənt/	*n.* 娱乐；消遣	
26	ultimately /ˈʌltɪmətli/	*adv.* 最后；根本	
27	optimize /ˈɒptɪmaɪz/	*v.* 最优化，完善	
28	utility /juːˈtɪləti/	*n.* 实用；效用	
29	the Neolithic Age	新石器时代	
30	the pipa	琵琶	
31	the flute	笛子	
32	the zither	扁琴	
33	*Li* (courtesy)	礼法	
34	*Yue* (music)	乐舞	
35	*She* (archery)	射箭	
36	*Yu* (chariot-riding)	驾车	
37	*Shu* (calligraphy or writing)	书法	
38	*Shu* (computation or mathematics)	算术	

Task 2 Comprehension

1 Translate the following paragraph about guqin into Chinese.

Guqin has been the most elegant traditional Chinese musical instrument and the most popular means of self-cultivation by scholars since it was invented by Fu Xi, or by Huang Di over 3,000 years ago. Originally it was played in grand worship ceremonies, later welcomed by all knowledgeable people in Chinese history.

2 Draw a mind map of the development of traditional Chinese music according to the passage.

The Development of Traditional Chinese Music

Section V
Insight into Proverbs

"Zhong" 忠—Loyalty

The themes of Peking Opera mainly focus on loyalty: fidelity, upright, and forbearance, and as "loyalty", seeking goodness from the heart and conscientiousness from the outside, is a kind of fine moral character advocated by the traditional Chinese culture. Since ancient times, the Chinese nation has a fine tradition of serving the country faithfully and sacrificing one's life for righteousness. "Every man is responsible for the rise and fall of the world." This is the common aspiration of all generations of Chinese people with lofty ideals. To be loyal to the cause, to the motherland and to the people is the highest and most sacred value pursuit of the Chinese nation.

1. 英雄所见略同。

Great heroes generally see things the same way.

(Great minds think alike.)

2. 君子之交淡如水; 小人之交甘若醴。

The friendship between gentlemen appears indifferent but is pure like water; the friendship between mean men is as sweet as wine.

3. 宁为玉碎, 不为瓦全。

Better to be a broken piece of jade than an unbroken shard of clay tile.

(Better to die with honor than live with dishonor.)

4. 真金不怕火炼。

Real gold does not fear the furnace.

(A person's true character is revealed in adversity.)

5. 天网恢恢，疏而不漏。

Heaven's net is vast; it's cast far and wide, but lets nothing through.

(No one escapes justice.)

6. 良药苦口利于病，忠言逆耳利于行。

Good medicine may be bitter in the mouth, but will help you recover; honest advice may offend the ear, but will aid your conduct.

7. 一失足成千古恨。

One false step can cause life-long regret.

8. 人为财死，鸟为食亡。

People die in pursuit of wealth; birds die in pursuit of food.

(Greed can lead to one's demise.)

9. 人无忠信，不可立于世。

A man without faithfulness cannot stand in the world.

10. 救人救到底，送人送到家。

When you save someone, save them all the way; when you see someone off, see them all the way home.

(Anything worth doing is worth doing right.)

11. 宝剑赠与烈士，红粉赠与佳人。

Give a treasured sword to a brave warrior; give rouge to a beautiful woman.

(Gifts should suit the person to whom they're given.)

12. 忠诚敦厚，人之根基也。

Loyalty and generosity are the foundation of human beings.

13. 君子以忠。

A wise man takes loyalty as the foundation.

14. 尽心于人曰忠，不欺于己曰信。

To be dedicated to others is for loyalty, not to deceive yourself is for faith.

15. 烈火验真金，艰难磨意志。

The fire is the test of gold, adversity of strong man.

Section VI
Self-assessment Checklist

1 Now, it's time for you to review your performance after learning this unit. Carry out a self-assessment by checking the following table.

Items	Ratings			
1. Knowledge	A	B	C	D
I know the origin and the essence of Peking Opera and Chinese music.				
I know the features of Peking Opera and the functions of Chinese music.				
I know the color culture in masks and doctrines of learning Chinese music.				
I know the etiquette of Peking Opera on stage.				
I master the expressions of Peking Opera and Chinese music.				
2. Skills	A	B	C	D
I can guess the new words through compound word-formation.				
I can identify "opinion" and "fact" in reading.				
I can use the method of scanning to get information.				
3. Speaking	A	B	C	D
I can talk Peking Opera or Chinese music fluently.				
I can explain the reasons of color culture in face masks logically.				
I can tell some stories related to loyalty in Chinese history.				
I can sing some Hongge (revolutionary songs) proudly.				
4. Confidence in Chinese Culture	A	B	C	D
I can appreciate Peking Opera.				
I can think critically some traditional etiquette in Peking Opera.				
I can respect other cultures.				
I can integrate traditional Chinese art culture with the western one.				

A: Basically agree

B: Agree

C: Strongly agree

D: Disagree

2　Fill in the blanks in the pictures below to check whether you have a good understanding of this unit.

(1) Write the role's name in Peking Opera.

(2) Make a mind map of one-day tour of Beijing culture in groups.

Direction: The one-day tour should include tangible and intangible culture heritages in Beijing. Tangible culture heritage involves the traditional landmarks or attractions like the Great Wall, the Forbidden City, Hutong (Beijing Courtyard), etc., while intangible culture heritage refers to all forms of social customs and habits, folklore, performing arts, rituals, festivals, traditional crafts and various knowledge and practice about nature and universe like food, Peking Opera, etc. in Beijing. The journey starts from your hotel.

Unit 7

Traditional Chinese Clothing

Traditional Chinese costumes, known as the quintessence of Chinese culture, are the precious wealth created by the Chinese nation. As a multi-ethnic country with a long history and splendid culture, Chinese clothing shows various styles. On formal occasions, we often see that many big figures or celebrities wear Cheongsam or Zhongshanzhuang, which are the typical symbols of traditional Chinese clothing. In recent years, there has been a revival of traditional Chinese attires. Some young people begin to engage in promoting Hanfu through the Internet social networking.

Being put in the first place in "yi, shi, zhu, xing" — clothing, food, shelter and means of travel, clothing has always played an important role in the necessity of people's life. So learning traditional Chinese clothing and its arts is another door to know the Chinese culture, helping further understanding the values, aesthetics, and etiquette of the country.

Overview

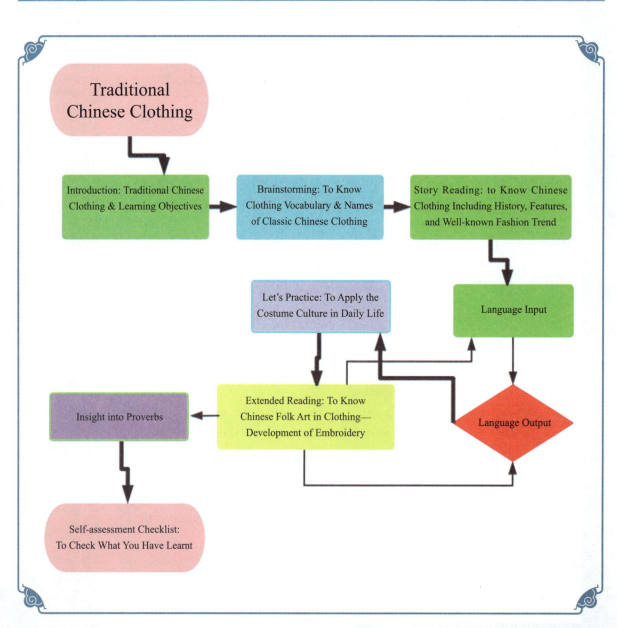

Traditional Chinese Clothing

Introduction: Traditional Chinese Clothing & Learning Objectives

Brainstorming: To Know Clothing Vocabulary & Names of Classic Chinese Clothing

Story Reading: to Know Chinese Clothing Including History, Features, and Well-known Fashion Trend

Language Input

Let's Practice: To Apply the Costume Culture in Daily Life

Extended Reading: To Know Chinese Folk Art in Clothing— Development of Embroidery

Insight into Proverbs

Language Output

Self-assessment Checklist: To Check What You Have Learnt

Learning Objectives

After learning this unit, students are able to reach the goals below.

1	专业能力目标	① 了解中国传统服饰、刺绣的演变过程及其在社会礼仪中的意义
		② 掌握与服饰文化相关的词汇、句型表达
		③ 提升阅读速度，把握文章的主旨和大意；能流利地讲述传统服饰及刺绣的发展、特点及其流行元素对世界时尚的影响
2	方法能力目标	① 运用表达时间的副词有条理地讲述某一事件的发展过程
		② 掌握连词或副词，如：although, however, as在行文中的意义
		③ 能运用scanning获取服饰和刺绣之间的联系
		④ 学会在阅读中通过上下文理解词义，扩大同义词库
3	社会能力目标	① 通过主题阅读，拓展知识面；运用与服饰相关的语言传达思想，服务社会和企业
		② 通过网络了解时尚文化，提高服饰的审美能力和相关批判性思维能力
		③ 掌握相关的英语阅读技巧，提高获取重要信息的能力
		④ 能运用恰当的服饰语言与基本知识传播中国经典的服饰文化及其礼仪
4	情感与思政目标	① 通过本单元学习，提升对中国传统服饰的认识；能正确看待中国传统服饰对国际时尚潮流的影响；增强责任感和使命感
		② 通过主题阅读，了解服饰文化的表现形式和内涵；提升对多元文化艺术的理解；激发传播中国服饰文化的热情；具备传播中国服饰文化的能力
		③ 通过主题及拓展阅读，提升着装方面的社交礼仪素养，塑造正确的价值取向和服饰消费观

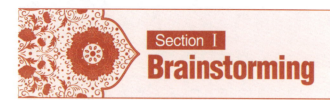

Section I
Brainstorming

Task 1   **Let's Warm Up**

1   Brainstorm the words you know about clothing.

2 Look at the following pictures and match them with their English expressions.

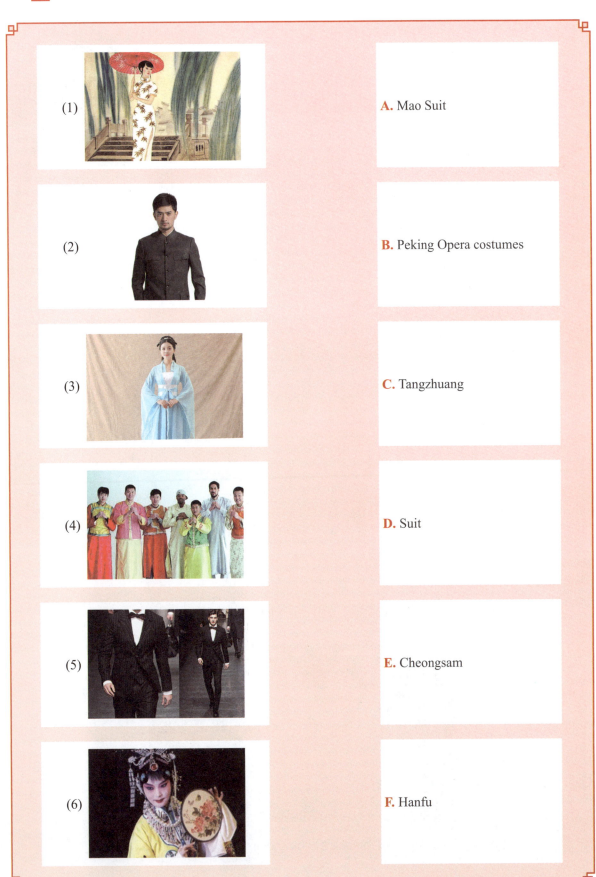

(1)

A. Mao Suit

(2)

B. Peking Opera costumes

(3)

C. Tangzhuang

(4)

D. Suit

(5)

E. Cheongsam

(6)

F. Hanfu

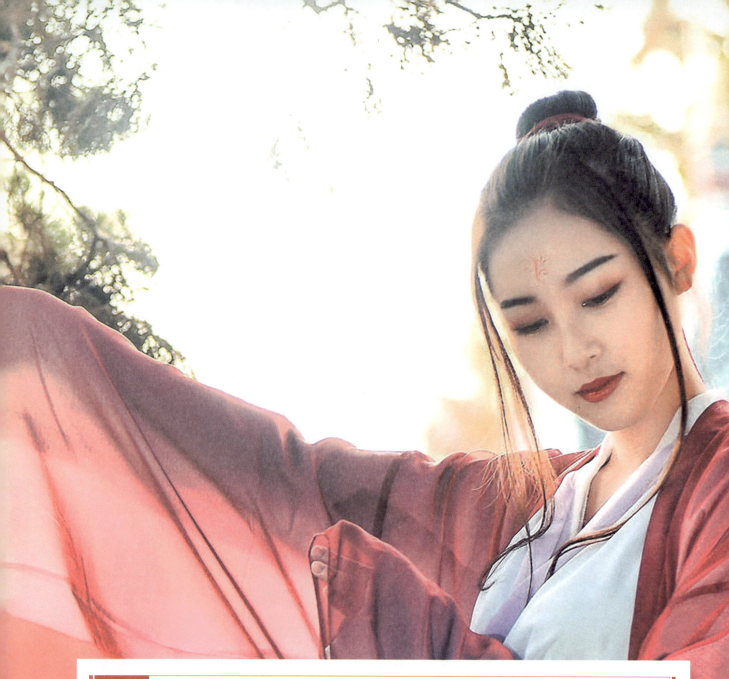

Task 2   Let's Watch and Say

1   Watch the video and see whether you can identify those attire people wear.

2   Talk about the influence of Chinese clothing elements in modern fashion.

Section II
Story Reading

Task 1　**Let's Read**

Chinese Clothing

As a vital part of Chinese civilization, traditional clothing plays an important role in the country's history and culture. The basic features are cross-collar, wrapping the right lapel over the left, tying with sash and a form of blouse plus skirt or long gown. These features had been preserved for thousands of years till 1930s, when Chinese Tunic Suit (Mao Suit/Zhongshanzhuang) and cheongsam prevailed. Nowadays, however, most Chinese wear modern clothes in their daily lives, not much difference from their western counterparts. Traditional attire is only worn during certain festivals, ceremonies or religious occasions and is often seen in Chinese television serials and movies. Many ethnic minorities wear their traditional costumes in their daily lives.

1. History

Based on historical discoveries, Chinese clothes date back to the later era of Paleolithic Times (1.7 million years ago—the 21st century BC). Materials used were of animal skins and decorations were of small stones and animal teeth. The "real" clothes were not invented until about 5,000 years ago by the Yellow Emperor (Huang Di). By the Shang Dynasty (17th century BC—1046 BC), the basic features of traditional Chinese attire were created, as well as the general pattern of blouse plus skirt. Later, the long gown appeared during the Zhou Dynasty (1046 BC—256 BC) and it co-existed with the blouse-skirt combinations for thousands of years, improving further as time passed. Then a great change occurred in the 1920s, when Mao Suit became popular among males and cheongsam among females. In the early period of the People's Republic of China, Mao Suit stayed popular among not only males, but also females. Later in the 1970's, when the country implemented reform and opening up policy, the masses gradually turned to western-style attire.

2. Features

In addition to the basic features and patterns, traditional Chinese attire has many other features like appearance, cutting, decoration, color and design, etc., all of which changed over the various dynasties. For example, black is the most dignified color in the Xia Dynasty (21st century BC—17th century BC), white in the Shang Dynasty and red in the Zhou Dynasty. They also varied based on one's political position, social status, occupation and gender, etc. For instance, dragon embroideries and bright yellow can only be used by emperors most of the time; in the Tang Dynasty (618—907), purple official costumes are for the fifth or higher rank officials; in the Qing Dynasty (1644—1911), the higher a person's social rank or the richer one was, the more embroideries and borders there were on his attire.

3. Well-known Fashion

Although the fashion trend changes over time, there are several types that are popular till today both at home and abroad.

● Traditional Han Chinese Clothing (Hanfu): It refers to the attire worn by the Han people from the enthronement of the Yellow Emperor (about 2698 BC) till the late Ming Dynasty (1368—1644). It became known as the Hanfu ("fu" means "clothes" in Chinese) because the fashion was improved and popularized during the Han Dynasty. It is usually in the form of long gown, cross collar, wrapping the right lapel over the left, loose wide sleeves and no buttons but a sash. Although simple in design, it gives different feelings to different wearers.

● Chinese Tang Suit (Tangzhuang): It is a combination of the Manchu male jacket of the Qing Dynasty and the western style suit. It is usually straight collared, with coiled buttons down the front. Its color and design are in traditional Chinese style but tailoring is western.

● Cheongsam (Qipao): Originated from the Manchu female clothes, it evolved by merging with western patterns that show off the beauty of a female body. Its features are straight collar, strain on the waist, coiled buttons and slits on both sides of the dress. Materials used are usually silk, cotton and linen. Cheongsam is the most popular Chinese attire in the world today.

● Chinese Tunic Suit (Mao Suit/Zhongshanzhuang): Also called the Sun Yat-sen Suit, it is designed by Dr. Sun Yat-sen by combining the western-style suit and Chinese attire. It has a turn-down collar and four pockets with flaps. As Chairman Mao Zedong worn it quite frequently, it is also called the Mao Suit by westerners. It is the main attire from the founding of the People's Republic of China in 1949 till 1980's. The country's leaders still wear it today when attending important occasions, such as military parades.
(836 words)

Task 2 Let's Learn

1 Read the following words and expressions, and tick the ones you know in the last column of the word list.

Numbers	Words & Expressions	Meanings	Notes
1	sash /sæʃ/	n. 腰带;肩带;饰带	
2	blouse /blaʊz/	n. 宽松的上衣;女装衬衫	
3	gown /gaʊn/	n. 长袍,长外衣;礼服;睡袍	
4	preserve /prɪˈzɜːv/	vt. 保存;保护	
5	prevail /prɪˈveɪl/	vi. 盛行,流行	
6	counterpart /ˈkaʊntəpɑːt/	n. 副本;配对物;极相似的人或物	
7	attire /əˈtaɪə(r)/	n. 服装;盛装	
8	ceremony /ˈserəməni/	n. 典礼,仪式	
9	occasion /əˈkeɪʒn/	n. 时机,机会;场合	
10	ethnic /ˈeθnɪk/	adj. 种族的;人种的	
11	minority /maɪˈnɒrəti/	n. 少数民族;少数派;未成年	
12	costume /ˈkɒstjuːm/	n. 服装,装束,戏装	
13	decoration /ˌdekəˈreɪʃn/	n. 装饰,装潢	
14	co-exist /kəʊ ɪgˈzɪst/	v. 共存	
15	combination /ˌkɒmbɪˈneɪʃn/	n. 结合;组合	
16	implement /ˈɪmplɪm(ə)nt/	vt. 实施,执行;实现	
17	mass /mæs/	n. 块,团;群众	
18	dignify /ˈdɪgnɪfaɪ/	vt. 使高贵;增威严;授以荣誉	
19	status /ˈsteɪtəs/	n. 地位;状态;重要身份	
20	occupation /ˌɒkjuˈpeɪʃn/	n. 职业;占有	
21	gender /ˈdʒendə(r)/	n. 性别;(用于某些语言)性(阳性、阴性和中性,不同的性有不同的词尾等)	
22	embroidery /ɪmˈbrɔɪdəri/	n. 刺绣;刺绣品	
23	enthronement /ɪnˈθrəʊnmənt/	n. 继位;登基典礼	
24	evolve /ɪˈvɒlv/	vt. 发展;进化	
25	merge /ˈmɜːdʒ/	vt. 合并;使合并;吞没 vi. 合并;融合	
26	frequently /ˈfriːkwəntli/	adv. 频繁地,经常地	
27	parade /pəˈreɪd/	n. 游行;阅兵	
28	a form of	一种……的形式	
29	as well as	也,和……一样	
30	based on	以……为基础,基于……	
31	to date back to	追溯到,从……开始有	
32	in addition to	除……外(还有)	
33	at home and abroad	国内外	
34	refer to	参考,涉及;指的是	
35	the reform and opening policy	改革开放政策	

2 Read and learn the key terms of traditional Chinese clothing.

Numbers	Terms	Meanings	Notes
1	the People's Republic of China	中华人民共和国	
2	Paleolithic Times	旧石器时代	
3	Hanfu	汉服	
4	the Yellow Emperor	黄帝	
5	long gown	长袍	
6	cross collar	一字领，交叉领	
7	to wrap the right lapel over the left	把右领搭在左领边	
8	to loose wide sleeves	宽松袖子	
9	no buttons but a sash	没有扣子，只有腰带	
10	Chinese Tunic Suit (Mao Suit/ Zhongshanzhuang)	中山装	
11	turn-down collar	翻领	
12	pockets with flaps	带襟翼的口袋	
13	Cheongsam (Qipao)	旗袍	
14	to strain on the waist	腰部收紧	
15	slits on both sides	两边开叉	
16	Chinese Tang suit (Tangzhuang)	唐装	
17	straight collar	立领	
18	coiled buttons down the front	前襟盘钮	

3 Get to know some useful expressions for etiquette of traditional Chinese clothing.

(1) Dress code: Dressing has always been the main topic in people's chat. Representing one's social status, Chinese people have attached much importance to their clothes. They think wearing decently is the best way to show their respecting to others and esteeming to themselves. Different occasions have different dress codes.

Dress Code / Occasions	Dos Patterns	Dos Colors	Don'ts Patterns	Don'ts Colors
Weddings	suit, dress	bright color: red	weird, body-revealed	dark color
Funerals	gown, pants	dark color: black	low-cut collar tights	bright color: red
Business	formal suit	mild color: blue	slippers vests	bright color
Office	formal or leisure		shorts button-open	

(2) Critical thinking: Supposing you are invited to attend a birthday party of your Chinese friend at his home with his family, what would you like to wear for the party?

Task 3 Consolidated Exercises

1 Decide whether the statements are True or False.

(1) People often wear traditional clothes in their daily life.

(2) The early Chinese clothes were made of animal skins with decorations of stones and animal teeth.

(3) The "real" clothes came into being before 5,000 years ago.

(4) Chinese traditional clothes began to change a lot in the founding of New China.

(5) Hanfu, created by the Yellow Emperor, was improved and popularized during Han Dynasty.

(6) The basic features of traditional Chinese clothes were different colors.

(7) If a person wore clothes with little or no embroideries on them, it meant that he/she was from a lower class or poor family.

(8) Chinese suit combines the jacket of Manchu male in the Ming Dynasty with the western style.

(9) The cutting of Cheongsam can show off the beauty of a female figure.

(10) Chinese Tunic Suit is also called Mao Suit because it was designed by Sun Yat-sen.

2 Fill in the brackets with the proper words given.

(1) We will do everything _____ (preserve) peace.

(2) Hanfu has been _____ (prevail) in the world in recent years.

(3) Chinese people like to _____ (decoration) their houses with paper cutting in Spring Festival.

(4) It is said that Cheongsam _____ (evolve) from Manchu female clothes.

(5) The origin of traditional Chinese clothing can _____ (date back to) Paleolithic Times.

(6) He _____ (frequent) donates large sums to charity.

(7) A _____ (combine) of talent, hard work and good looks have taken her to the top.

(8) It's best not to try a new recipe for the first time on such an important _____ (occasion).

3 Translate the following paragraph into English.

作为中华文明的重要组成部分, 传统服饰在中国历史和文化中扮演着重要的角色。一直以来, 它是礼仪的重要体现。中国人认为着装既代表了身份和地位, 得体的着装更是代表了一个人的修养, 表示对他人和对自己的尊重。

Section III
Let's Practice

Task 1 **Let's Talk**

1 Watch the video to know how to wear Hanfu. Try to write down the following points.

Blouse: _____(1)_____

Skirt: Move on to the skirt. _____(2)_____

First step into the center of the skirt. _____(3)_____ .

_____(4)_____ . Now bring up the front panel,

wrap around the back and tie it again in front. To

create more friction, I loop the remaining bound

around each side several more times.

Hezi: Turn it to the back, _____(5)_____ ,

then the outer set of ties. _____(6)_____ .

Sleeves Coat: _____(7)_____ . Tie them together at the bottom.

Long scarf (Pibo): I'm also adding _____(8)_____ Pibo, draped around the arms for

decoration.

2 Finish the outline of the features of traditional Chinese clothing.

Clothes Style	Features				
	Pattern	Collar	Button	Sleeve	Pocket
Hanfu	long gown				
Cheongsam					
Chinese Tang Suit					
Chinese Tunic Suit					

Task 2 Team Work

1 Presentation.

Each team chooses one type of the traditional Chinese clothes, and make an introduction of it.

2 Translation and discussion.

Translate the Chinese saying "人靠衣装,马靠鞍" into English, and then answer the following questions.

(1) What is the original story?

(2) How do you understand it?

(3) Do you think it right or not?

3 Critical thinking.

(1) Why does Chinese leader often wear Chinese Tunic suit in state visit? Why is Chinese Tunic Suit called Mao Suits or Zhongshanzhuang?

(2) As we know, the features of Hanfu are the cross collar and loose wide sleeves, then what are the implications of these features?

Chinese Embroidery Development

Embroidery is a brilliant pearl in Chinese art. From the magnificent Dragon Robe worn by Emperors to the popular embroidery seen in today's fashions, it adds so much pleasure to Chinese people's lives and their culture.

The oldest embroidered product in China on record dates from the Shang Dynasty. Embroidery in this period symbolized social status. It was not until later on, as the national economy developed, that embroidered products entered the lives of the common people.

The Han Dynasty witnessed a leap in embroidery in both technique and art style. Court embroidery was set and specialization came into being. The patterns covered a larger range, from sun, moon, stars, mountains, dragons, and phoenix to tiger, flower and grass, clouds and geometric patterns. Auspicious words were also fashionable.

During the Three Kingdoms Period, one notable figure in the development of embroidery was the wife of Sun Quan, King of Wu. She was also the first female painter recorded in Chinese painting history. She was good at calligraphy, painting and embroidery. Sun Quan wanted a map of China and she drew one for him and even presented him embroidered map of China. She was reputed as the Master of Weaving, Needle and Silk. Portraits also appeared on embroidered things during this time.

As Buddhism boomed in China during the Wei, Jin, Sui and Tang dynasties, embroidery was widely used to show honor to Buddha statues. Lu Meiniang, a court maiden in the Tang Dynasty, embroidered seven chapters of Buddhist Sutra (法华经七卷) on a tiny piece of silk. New skill in stitching emerged during this period. Besides Buddhist figures, the subjects of Chinese painting such as mountains, waters, flowers, birds, pavilions and people all became themes of embroidery, making it into a unique art.

The Song Dynasty saw a peak of development of embroidery in both quantity and quality. It developed into an art by combining calligraphy and painting. New tools and skills were invented. The Wenxiu Department was in charge of embroidery in the Song court. During the reign of Emperor Huizong, the patterns of embroidery was divided into four categories: mountains and waters, pavilions, people, and flower and birds, and two functions: art for daily use and art for art's sake.

The religious related to embroidery was strengthened by the rulers of Yuan Dynasty. Embroidery was much more applied in Buddha statues, sutras and prayer flags. One product of this time is kept in Potala Palace.

As the sprout of capitalism emerged in Ming Dynasty, Chinese society saw a substantial flourish in many industries. Embroidery showed new features, too. Traditional auspicious patterns were widely used to symbolize popular themes: Mandarin ducks for love; pomegranates for fertility; pines, bamboos and plums for integrity; peonies for riches and honor; and cranes for longevity. The famous Gu Embroidery is typical of this time.

The Qing Dynasty inherited the features of the Ming Dynasty and absorbed new ingredients from Japanese embroidery and even Western art. New materials such as gilded cobber and silvery threads emerged. According to *The Dream of the Red Chamber*, a popular Chinese novel set during the Qing Dynasty, peacock feathers were also used.

Along the history, four distinctive schools of Chinese embroidery gradually come into being: Su embroidery, Shu embroidery, Xiang embroidery, and Yue embroidery, which are now designated by the government as a Chinese Intangible Cultural Heritage. Some ethnic groups like Miao also have their own embroidery. Each of the embroidery reached their summit after the blossoming of the Silk Road trade created a demand for Chinese goods. (585 words)

Task 1　New Words and Expressions

Read the following words and expressions, and tick the ones you know in the last column of the word list.

Numbers	Words & Expressions	Meanings	Notes
1	brilliant /ˈbrɪliənt/	*adj.* 灿烂的，闪耀的；杰出的	
2	pearl /pɜːl/	*n.* 珍珠	
3	magnificent /mæɡˈnɪfɪsnt/	*adj.* 高尚的；壮丽的	
4	symbolize /ˈsɪmbəlaɪz/	*vt.* 象征	
5	witness /ˈwɪtnəs/	*n.* 证人；目击者	
6	specialization /ˌspeʃəlaɪˈzeɪʃn/	*n.* 专门化；特殊化	
7	range /reɪndʒ/	*n.* 范围；幅度	
8	phoenix /ˈfiːnɪks/	*n.* 凤凰；死而复生的人	
9	geometric /ˌdʒiːəˈmetrɪk/	*adj.* 几何学的	
10	auspicious /ɔːˈspɪʃəs/	*adj.* 有利的；吉祥的	
11	notable /ˈnəʊtəbl/	*adj.* 显著的；著名的	
12	portrait /ˈpɔːtreɪt/	*n.* 肖像；描写	
13	Buddhism /ˈbʊdɪzəm/	*n.* 佛教	
14	boom /buːm/	*n.* 繁荣　*v.* 使兴旺	
15	statue /ˈstætʃuː/	*n.* 雕像，塑像	
16	maiden /ˈmeɪdn/	*n.* 少女，处女	
17	sutra /ˈsuːtrə/	*n.* 佛经（等于sutta）；箴言	
18	stitch /stɪtʃ/	*n.* 针脚，线迹；一针 *v.* 缝，缝合	
19	peak /piːk/	*n.* 山峰；最高点	
20	sprout /spraʊt/	*n.* 芽；萌芽　*vi.* 发芽；长芽	

Numbers	Words & Expressions	Meanings	Notes
21	substantial /səbˈstænʃl/	*adj.* 大量的；实质的	
22	Mandarin duck	鸳鸯	
23	pomegranate /ˈpɒmɪɡrænɪt/	*n.* 石榴	
24	fertility /fəˈtɪləti/	*n.* 多产；肥沃	
25	plum /plʌm/	*n.* 李子；梅子	
26	integrity /ɪnˈteɡrəti/	*n.* 完整；正直；诚实	
27	peony /ˈpiːəni/	*n.* 牡丹；芍药	
28	crane /kreɪn/	*n.* 鹤	
29	ingredient /ɪnˈɡriːdiənt/	*n.* 原料；要素；组成部分	
30	gilded cobber	镀金的	
31	peacock /ˈpiːkɒk/	*n.* 孔雀	
32	designate /ˈdezɪɡneɪt/	*vt.* 指定；指派	
33	summit /ˈsʌmɪt/	*n.* 顶点；最高级会议	
34	blossom /ˈblɒsəm/	*n.* 花；开花期；兴旺期 *v.* 开花；兴旺	

Task 2 Comprehension

Fill in the table below about the features of each period of the embroidery after reading the passage.

Periods	Features of embroidery
The Shang Dynasty	Symbolized social status, later on that embroidered products entered the lives of the common people
The Han Dynasty	
Three Kingdoms Period	
During the Wei, Jin, Sui and Tang Dynasties	
The Song Dynasty	
The Yuan Dynasty	
The Ming Dynasty	
The Qing Dynasty	
Modern Times	Four distinctive schools are now designated by the government as a Chinese Intangible Cultural Heritage

Insight into Proverbs

"Xiao" 孝—Filial Piety

Filial piety is the foundation of morality and is the first priority in all good deeds. It is of great significance to carry forward and rebuild the new filial piety adapted to modern civilized society in building a harmonious socialist society.

1. 万恶淫为首，百善孝为先。

Of the myriad sins, lewdness heads the list; of the many virtues, filial piety is the first.

2. 当家才知柴米贵；养子方晓父母恩。

You have to run your own household to know the price of rice and firewood. You have to raise your own children to understand the sacrifices that parents make.

(It's only when we have children of our own that we come to understand how much is required of parents.)

3. 子孝父心宽。

If the son is filial, his father's heart can rest easy.

4. 子不嫌母丑，狗不嫌家贫。

A son never minds an ugly mother, and a dog never minds a poor house.

(Dogs don't dislike their mother's ugliness or their family's poverty.)

5. 父母之年，不可不知也。一则以喜，一则以惧。

A son should always keep in mind the age of his parents, as a matter of thankfulness as well as for anxiety.

6. 弟子入则孝，出则弟。

A young man, when at home, should be a good son; when out in the world, a good citizen.

7. 事父母，能竭其力。

A man in his duties to his parents is ready to do his utmost.

8. 父母之恩，水不能溺，火不能灭。

The grace of parents is something that water can not drown and fire can not extinguish.

9. 父母在，不远游，游必有方。

While his parents are living, a son should not go abroad far away; if he does, he should let them know where he goes.

10. 要如亲恩，看你儿郎；要求子顺，先孝爹娘。

If you want to be kind, see your son; ask for a filial son, first be filial to your father and mother.

11. 不得乎亲，不可以为人；不顺乎亲，不可以为子。

You can't be a man unless you are close to parents or a son unless you are close to parents.

12. 父母天地心，大小无厚薄。

Parents' hearts and minds are not thick to one or thin to another.

13. 为人子，止于孝；为人夫，止于慈。

To be a son is to be filial; to be a husband is to be kind.

14. 树欲静而风不止，子欲养而亲不待。

Trees want to be quiet but wind won't, sons want to be filial to parents who can't wait.

(Trees prefer calm while wind does not subside; Sons choose filial while parents died.)

15. 诗书立业，孝悌做人。

Poetry and books make a career and filial piety makes a man.

Self-assessment Checklist

1 Now, it's time for you to review your performance after learning this unit. Carry out a self-assessment by checking the following table.

Items	Ratings			
1. Knowledge	A	B	C	D
I know the development of Chinese clothing and embroidery.				
I know the features of different style of clothes and embroidery in different dynasties.				
I know the basic dress code in different occasions.				
I know the features of different style of clothes and embroidery in different dynasties.				
I master the relationship between Chinese costumes and embroidery.				
I master the expressions of traditional Chinese clothing and embroidery.				
2. Skills	A	B	C	D
I can use adverbs of time to state an event in chronological way.				
I master the conjunction or adverb "although", "however" in reading.				
I can read through context to expand synonym.				
I can apply dress code in communication.				
3. Speaking	A	B	C	D
I can talk one of the Chinese costume styles fluently.				
I can explain the reasons of some Chinese sayings about costumes logically.				
I can state the development of Chinese embroidery fluently.				
I can tell the connection of traditional Chinese clothing and embroidery.				
4. Confidence in Chinese Culture	A	B	C	D
I understand the essence of Chinese folk arts.				
I can dress properly in different occasions.				
I master aesthetics of dressing Chinese style clothing.				
I am proud of traditional Chinese costume.				

A: Basically agree

B: Agree

C: Strongly agree

D: Disagree

2 Fill in the blanks in the pictures below to check whether you have a good understanding of this unit.

(1) Try to name the parts of the clothing below.

Hanfu

Mao Suit

Cheongsam(Qipao)

(2) Mind map of the etiquette of Chinese costumes.

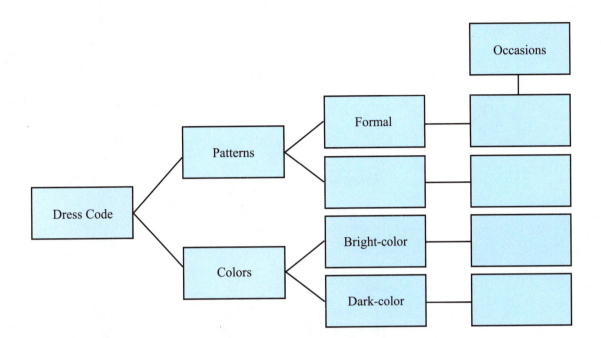

捌

Unit 8

Chinese Food Culture

China is a vast country with diverse climates, customs, products, and habits. People living in different regions display great variety in their diets. As a result, foods vary from north to south, and tastes also differ regionally. Generally speaking, there are three essential factors by which Chinese cooking is judged, namely "color, aroma, and taste". "Color" refers to the layout and design of the dishes. "Aroma" implies not only the smell of the dish, but also the freshness of the materials and the blending of seasonings. "Taste" involves proper seasoning and fine slicing techniques. These three essential factors are achieved by careful coordination of a series of delicate activities: selecting ingredients, mixing flavors, timing and cooking, adjustment of the heat, and laying out the food on the plate.

Chinese food culture is extensive and profound. It not only embodies the three essential factors in making a dish, but also in its table etiquette, serving tea and drinking wine by playing a drinkers' wager game during a banquet or a common gathering.

Overview

```
        Chinese
        Food Culture
            │
            ▼
        Introduction:
        Chinese Food & Learning
        Objectives
            │
            ▼
Brainstorming:
To Know the Vocabulary &
Some Names of Chinese
Dishes
            │
            ▼
        Story Reading:
        To Know Chinese Food Including
        History, Divisions, and Features.
            │
            ▼
Language Input:
Text reading—Vocabulary
Language Output:
Comprehension—Vocabulary—
Translation
            │
            ▼
        Let's Practice
        Chopsticks
            │
            ▼
Watch and Talk
Team Work
            │
            ▼
        Extended Reading:
        To Know Chinese Tea
        Culture
            │
            ▼
Vocabulary
Comprehension
            │
            ▼
        Insight into Proverbs
            │
            ▼
        Self-assessment
        Checklist:
        To Check What You
        Have Learnt
```

Learning Objectives

After learning this unit, students are able to reach the goals below.

1	专业能力目标	① 了解中国饮食文化中餐桌茶酒礼仪的意义
		② 掌握更多与饮食文化相关的词汇及句型表达
		③ 提高阅读速度,快速掌握文章的主旨和大意;能运用较流利的语言讲述中国经典菜系与茶艺、茶道的特点及内涵
2	方法能力目标	① 掌握同位语的解释功能,并能运用其提高阅读理解力
		② 掌握非限制性定语从句的功能,并能运用其提高阅读理解能力
		③ 掌握非谓语动词作为后置定语在阅读中的功能
		④ 运用寻读和略读的方法快速获取关键信息
3	社会能力目标	① 通过主题阅读,能运用餐桌礼仪进行社交活动;具有为社会、企业提供饮食文化方面语言服务的意识,并具备相应的能力
		② 通过网络了解食品资源问题,能辩证地分析实际问题
		③ 通过主题阅读,能运用恰当的基本知识和语言讲述中国独具特色的饮食文化和餐桌礼仪,并能言传身教
4	情感与思政目标	① 通过本单元的学习,了解中国饮食文化对阴阳五行哲学思想、儒家伦理道德观念、中医养生学说的影响;了解饮食审美风尚;具备继承和弘扬中国传统文化的自觉性
		② 通过主题阅读,加深对中国饮食文化源流、特点的认识;能正确传播餐桌礼仪;提升茶道艺术修养
		③ 通过对饮食文化和茶文化中礼仪规范的理解,达到人与人、人与社会的和谐相处;领悟勤俭节约的思想;通过主题及拓展阅读,能理解中国饮食文化、茶文化的内涵,提升自我修养,塑造科学的饮食观和食品消费观
		④ 提高食品安全意识,具备分辨网络上关于食品信息的能力;不信谣不传谣,维护社会正常秩序

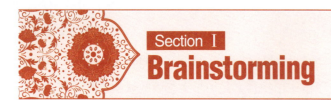

Section I
Brainstorming

Task 1 | **Let's Warm Up**

1 Brainstorm the words about the names of Chinese dishes you know.

2 Look at the pictures and match up the corresponding pairs.

(1) 宫保鸡丁

(2) 麻辣鸡

(3) 九转大肠

(4) 清蒸鱼

(5) 三杯鸡

(6) 糖醋里脊

A. Steamed Fish

B. Sweet and Sour Pork

C. Chicken with Chili and Sichuan Pepper

D. Kung Pao Chicken

E. Braised Intestines in Brown Sauce

F. Stewed Chicken with Three Cups Sauce

1 Watch the video and write down some names of Cantonese dim sum, and read them.

肠粉: _____

烧鹅: _____

虾饺: _____

烧卖: _____

招牌脆皮红米肠: _____

凤爪: _____

米粥: _____

叉烧包: _____

流沙包: _____

2 Think and speak out some famous snacks you've ever had in different places of China.

Section II
Story Reading

Task 1   Let's Read

Chinese Food

1. Chinese Food History

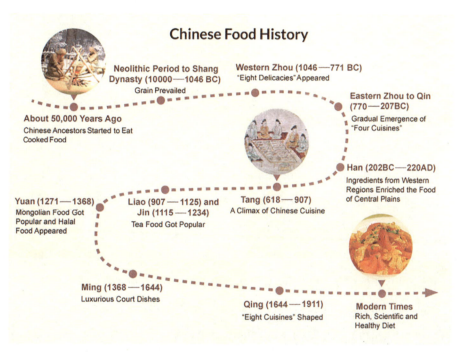

Chinese Food History

In the long Paleolithic Period (3 million years to 10 thousand years ago), human beings in the primitive society of China mainly lived on hunting. They didn't know how to use fire and ate everything raw. In that age there was no fine food and cooking. About 50,000 years ago, Chinese ancestor, Suiren, who was the head of the Three Sovereigns and Five Emperors, invented "manual wood drilling to make fire", which started the era of cooked food. By the Western Zhou Dynasty (1046 BC—771 BC), "Eight Delicacies" appeared, which marked that cooking had become an art. During the Eastern Zhou Dynasty to Qin Dynasty (770 BC—207 BC), four major cuisines, Shandong Cuisine, Jiangsu Cuisine, Guangdong Cuisine and Sichuan Cuisine, were successively born. From the Han Dynasty (202 BC—220 AD) to the Tang Dynasty (618—907), people in the central plains gradually blended their food with the ingredients from the western regions and the flavor of the minorities in the northwest, making the food more abundant. During the Ming Dynasty (1368—1644) and Qing Dynasty (1644—1911), the "Eight Cuisines" gradually formed, each with distinct features.

2. China's Regional Cuisines—Types and Features

China's cooking styles and dietary preferences can be divided into many geographical areas, and each area has a distinct style of cooking. Generally, China's regional cuisines taste are divided into 5 regions as follows:

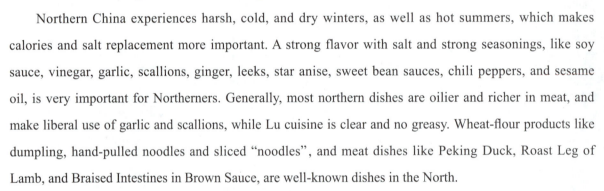

- Northern China food—salty, simple, less vegetables with wheat as the staple food.

  Regions: Beijing, Xi'an, Inner Mongolia, and Northeast China

- Western China food—hearty halal food with lamb the main meat

  Regions: Xinjiang, Tibet and Gansu

- Central China food—hot and spicy with a lot of seasonings

  Regions: Sichuan, Chongqing, Hunan

- Eastern China food—sweet and light

  Regions: Guangdong, Fujian, Zhejiang, Jiangsu, Anhui

- Southern minority food—sour with a lot of preserved ingredients

  Regions: Yunnan, Guizhou and Guangxi provinces

3. Flavors and Ingredients of Some Regional Cuisines

Northern China experiences harsh, cold, and dry winters, as well as hot summers, which makes calories and salt replacement more important. A strong flavor with salt and strong seasonings, like soy sauce, vinegar, garlic, scallions, ginger, leeks, star anise, sweet bean sauces, chili peppers, and sesame oil, is very important for Northerners. Generally, most northern dishes are oilier and richer in meat, and make liberal use of garlic and scallions, while Lu cuisine is clear and no greasy. Wheat-flour products like dumpling, hand-pulled noodles and sliced "noodles", and meat dishes like Peking Duck, Roast Leg of Lamb, and Braised Intestines in Brown Sauce, are well-known dishes in the North.

Eastern China cuisine includes five of the famous eight cuisines from the Pearl River Delta to the Yangtze River Delta. Food in the eastern China mainly features sweet and subtle flavor with fresh and natural ingredients from rivers, sea, and mountains, using sugar, wines, vinegar, and soy sauces and precise and delicate carving techniques. Their common cooking techniques include braising, stewing, sautéing, frying, steaming and boiling.

Cantonese cuisine tends to be mild. It is famous for its dim sum, bite-sized portions of food traditionally served in small steamer baskets or on small plates. Fujian cuisine is famous for its abundant ingredients from the sea and mountains. It is characterized by its fine slicing techniques, various soups and broths, and exquisite culinary art, like Buddha Jumping Wall (Sea Food and Poultry Casserole). Zhejiang cuisine comprises the styles of Hangzhou, Ningbo, Shaoxing, and Shanghai. It is famous for freshness, softness, and smoothness, with a mellow fragrance. It is characterized by its elaborate preparation. Jiangsu cuisine consists of Yangzhou, Nanjing, and Suzhou dishes. It is famous for its fresh taste, with moderate saltiness and sweetness.

The representative of Central China food including Hunan cuisine and Sichuan dishes mainly features hot and spicy with pickles, which is much more similar to that of minority. Common cooking techniques include pickling, smoking, stewing, stir-frying, and braising, and pot-roasting.

China is large and the climate, ingredients, history and dining customs vary from place to place, leading to the differences in cooking methods and dish flavors and forming the different cuisines. Chinese food has become one of the main stream among the world's three cuisine styles. (700 words)

Task 2　Let's Learn

1　Read the following words and expressions, and tick the ones you know in the last column of the word list.

Numbers	Words & Expressions	Meanings	Notes
1	primitive /ˈprɪmətɪv/	*adj.* 原始的，远古的	
2	blend /blend/	*v.* 混合；协调	
3	ingredient /ɪnˈɡriːdiənt/	*n.* 原料；要素	
4	flavor /ˈfleɪvə/	*n.* 情味，风味	
5	cuisine /kwɪˈziːn/	*n.* 烹饪，烹调法	
6	abundant /əˈbʌndənt/	*adj.* 丰富的；充裕的	
7	distinct /dɪˈstɪŋkt/	*adj.* 明显的；独特的	
8	feature /ˈfiːtʃə(r)/	*n.* 特色，特征	
9	dietary /ˈdaɪətəri/	*adj.* 饮食的，饭食的	
10	preference /ˈprefrəns/	*n.* 偏爱，倾向；优先权	
11	staple /ˈsteɪpl/	*n.* 主要产品；主食	
12	spicy /ˈspaɪsi/	*adj.* 辛辣的；香的	
13	seasoning /ˈsiːzənɪŋ/	*n.* 调味品，佐料	
14	preserve /prɪˈzɜːv/	*vt.* 保存；保护	
15	harsh /hɑːʃ/	*adj.* 严厉的；严酷的	
16	vinegar /ˈvɪnɪɡə(r)/	*n.* 醋	
17	garlic /ˈɡɑːlɪk/	*n.* 大蒜；蒜头	
18	scallion /ˈskæliən/	*n.* 青葱	
19	ginger /ˈdʒɪndʒə(r)/	*n.* 姜	
20	leek /liːk/	*n.* 韭菜	
21	anise /ˈænɪs/	*n.* 大茴香	
22	sauce /sɔːs/	*n.* 酱油；沙司；调味汁	
23	chili /ˈtʃɪli/	*n.* 红辣椒，辣椒	
24	sesame /ˈsesəmi/	*n.* 芝麻，胡麻	
25	greasy /ˈɡriːsi/	*adj.* 油腻的；含脂肪多的	
26	subtle /ˈsʌtl/	*adj.* 微妙的；精细的	
27	precise /prɪˈsaɪs/	*adj.* 精确的；明确的	

continued

Numbers	Words & Expressions	Meanings	Notes
28	delicate /ˈdelɪkət/	*adj.* 微妙的；精美的	
29	braise /breɪz/	*vt.* 炖；蒸	
30	stew /stjuː/	*vt.* 炖，煮，焖	
31	sauté /ˈsəʊteɪ/	*vt.* 嫩煎；炒	
32	slice /slaɪs/	*vt.* 切下；将……切成薄片	
33	broth /brɒθ/	*n.* 肉汤；高汤	
34	exquisite /ɪkˈskwɪzɪt/	*adj.* 精致的；细腻的	
35	culinary /ˈkʌlɪnəri/	*adj.* 烹饪的	
36	poultry /ˈpəʊltri/	*n.* 家禽	
37	casserole /ˈkæsərəʊl/	*n.* 勺皿；砂锅菜	
38	mellow /ˈmeləʊ/	*adj.* 圆润的；成熟的；芳醇的	
39	fragrance /ˈfreɪɡrəns/	*n.* 香味，芬芳	
40	elaborate /ɪˈlæb(ə)rət/	*adj.* 精心制作的	
41	moderate /ˈmɒdərət/	*adj.* 稳健的，温和的；适度的	
42	pickle /ˈpɪkl/	*n.* 泡菜；盐卤；腌制食品	
43	to make liberal use of	随意使用	
44	be famous for	因……而著名	
45	to consist of	由……组成	
46	be representative of	表示，以……为代表	

2 Read and learn the key terms of Chinese food, and tick the ones you know in the last column of the word list.

Numbers	Terms	Meanings	Notes
1	Paleolithic Times	旧石器时代	
2	the Three Sovereigns and Five Emperors	三皇五帝	
3	Suiren	燧人	
4	to drill wood to make fire	钻木取火	
5	Eight Delicacies	八珍	
6	the central plains	中原	
7	Eight Cuisines	八大菜系	
8	the Pearl River Delta	珠江三角地区	
9	the Yangtze River Delta	长江流域	
10	dim sum	点心	
11	Buddha Jumping Wall	佛跳墙	

3 Get to know some etiquette of Chinese dinning culture.

(1) Dinning manners: As an ancient civilization, China pays great attention to established etiquette. Actually, most table manners in China are similar to that of in the West. Table manner in China is very serious from seating arrangement to leaving the table. So when dinning in an formal banquet, you'd better follow some dinning etiquette and culture.

Table Manner

Items	Dos	Don'ts
Seating	Take a seat in accordance with the master of the banquet's arrangement.	Take a seat randomly.
Eating	1. Let older people eat first. 2. Pick up your bowl in proper ways. 3. Take food first from the plates in front of you. 4. Ask or consult others. 5. Concentrate on the meal and your companions. 6. Refill your bowl with rice yourself or help the elder. 7. Chew with closed mouth without making noises. 8. Slurp quietly. 9. Talk little and quietly. 10. Sneeze with hand or a handkerchief and turn away.	1. Steal a march on the elders. 2. Gobble food up. 3. Pick up too much food at a time. 4. Use your chopsticks to burrow through the food and "dig for treasure". 5. Keep your eyes glued to the plates. 6. Watch television, use your phone, or carry on some other activity. 7. Chew with opened mouth wide with loud noises. 8. Talk with others with your mouth full and loud. 9. Sneeze without hand or tissue.

(2) Some famous expressions for Chinese food culture.

A. 民以食为天，食以味为先。

Food is always taken as the first priority, while taste is the soul of food.

B. 每餐七分饱，健康活到老。

A half-full diet keeps you healthy until you are old.

C. 食以平衡阴阳。

Taking food properly is to balance Yin and Yang.

D. 食在广州，味在西关，厨出凤城。

Cuisines are in Guangzhou, snacks in Xiguan, and cooks from Fengcheng.

Task 3　Consolidated Exercises

1 Decide whether each statement is True or False according to the text.

(1) It was said that Suiren was the first person to make fire from drilling wood.

(2) About 50, 000 years ago Chinese people began to eat raw food.

(3) Eight Cuisines formed during Tang Dynasty.

(4) Northern dishes are oilier and richer in meat because of the weather in this area.

(5) Dim sum is one of the most popular Cantonese cuisine.

(6) Central China cuisine focuses on precise and delicate carving techniques.

(7) Southern minority food is spicy and hot with pickles.

(8) The reasons for forming different cuisines in China are mainly due to the climate, ingredients, history and dining customs.

2 List the vocabulary of ingredients and seasonings of making dishes from the passage.

3 Translate the following sentences into English.

(1) 这道菜肥而不腻，入口即化。

(2) 广式菜味道清淡，讲究保持食材的原味。

(3) 北方菜口味比较油腻，味道重。

(4) 南京菜融合了皇家的奢华和江南水乡的精致。

(5) 川菜以麻辣著称，主要分为成都和重庆两个系列。

Section III
Let's Practice

Task 1 Let's Talk

1 Think about the following questions and share your ideas.

(1) When did you begin to use chopsticks?

(2) What materials were used to make chopsticks?

(3) What is the proper way of using chopsticks?

(4) Do you know the culture of chopsticks in China?

2 Watch the video to know some culture about chopsticks.

(1) What theory does chopsticks apply to?

(2) What do chopsticks represent in Chinese culture?

Task 2 Team Work

1 Each team chooses one of the Chinese dishes, and present the making procedure and the features in an interactive way.

2 Discuss one of the sayings or famous expressions of Chinese food listed as follows.

(1) Food is always taken as the first priority, while taste is the soul of food.

(2) A half-full diet keeps you healthy until you are old.

(3) Taking food properly is to balance Yin and Yang.

(4) Cuisines are in Guangzhou, snacks in Xiguan, and cooks from Fengcheng.

Section IV
Extended Reading

Chinese Tea Culture

Home to China, tea has enjoyed a history of thousands of years in China, right back to the ancient Shennong period (some 5,000 years ago), and has played a significant role in Chinese art and society since ancient times. Tea in China, along with silk and porcelain, spread to the other part of the world through trading routes of Silk Road and Tea Horse Road.

Tea culture embodies the Chinese culture, not only including the level of material culture, but also spirit level. Chinese tea culture is extensive and profound, which can be traced back to Tang Dynasty. Since then, the spirit of tea had penetrated into the court and society and integrated into poem, painting, medicine, calligraphy and religion.

In the modern era, Chinese tea has become a symbol of culture and etiquette in all aspects of people's lives.

1. The Art of Chinese Tea Making

The art of making and serving tea are one of the ancient Chinese arts. In a teahouse, a girl wearing Hanfu or Cheongsam will make the tea for the guests. She pours some hot/boiled water into a cup, in which is put another even smaller cup, until it overflows a little. Then she fills a third one in which some tea leaves were placed, before setting the lid. Her gestures are accurate. She waits a little, and then decants finally the tea in the smallest of the cups. She asks the guests to turn the cup, and to smell. As the cup cools, the fragrance evolves to become milder. Drinking is now the next step subjected however to hold the cup with three fingers.

This particular art is popularly practised among the common people, be they Buddhists, Daoists or Confucianists, because tea is taken not just as a means of quenching thirst and ridding the body of excessive oil, but also to nurture the spirit—yi qing yang xing (怡情养性, to move the feelings and nurture the spirit).

2. The Customs of Tea Serving

For Showing Respect

Tea-drinking is a sign of respect in Chinese social life, whether in ancient or modern times. But the connotations are different.

In the past, Chinese people observed strict rules and rituals in serving tea. For example, people of lower social class had to show their respect to the upper classes by serving them a cup of tea on Spring Festival or at other festival times. Today, with the increasing liberalization of Chinese society, this rule has been suppressed by the prevailing custom to offer tea when someone comes for a visit, irrespective of class, showing the host's warm welcome and hospitality to the guest.

For Expressing Gratitude to Parents

It has been customary from ancient times for bride and groom to kneel before their parents and serve them tea, during the traditional Chinese wedding ceremony. This is a devout way of showing gratitude for the love and care received from parents ever since childhood. Sometimes the bride serves the groom's family, and the groom serves the bride's family. This process symbolizes the joining together of the two families.

When Offering Sacrifices to Ancestors and Gods

In China's colorful folk customs, there is a close relationship between "tea" and funeral sacrifices. The idea that "No tea means no mourning" is deeply rooted in Chinese rituals.

Using tea as a sacrifice, Chinese people worship the heaven, the earth, the god, Buddha and ghosts, and doing so is closely related to funeral customs. Tea is not the exclusive right of nobles, nor was it ever a royal patent; it is a common gift for the whole nation.

The ancient custom of offering tea as a sacrifice to ancestors and gods is well preserved even to this day, whether among the majority Han nationality or the ethnic minorities.

3. The Etiquette of Tea Drinking

Environment of Tea Drinking

The environment for drinking tea should be quiet, clean and comfortable, giving people the sense of being at ease.

The selected tea may vary according to individual. For instance, northerners generally like scented tea, but people from regions south of the Yangtze River love green tea, while Cantonese people prefer oolong tea.

Tea sets can either be beautifully crafted or simple and unadorned.

Don't pour the cup too full; 80% full would be fine.

"Thank you" Gesture in Tea Drinking

When someone pours tea into your cup, you can tap the table with your first two fingers two or three times, showing thanks to the pourer for the service and of being enough tea. The pourer will stop pouring when seeing the gesture. (792 words)

Task 1   New Words and Expressions

Read the following words and expressions, and tick the ones you know in the last column of the word list.

Numbers	Words & Expressions	Meanings	Notes
1	significant /sɪɡ'nɪfɪkənt/	*adj.* 重大的；有效的；有意义的	
2	embody /ɪm'bɒdi/	*vt.* 体现，使具体化	
3	penetrate /'penətreɪt/	*v.* 渗透，穿透	
4	integrate /'ɪntɪɡreɪt/	*v.* 使……完整	
5	discard /dɪ'skɑːd/	*v.* 抛弃；放弃	
6	accurate /'ækjərət/	*adj.* 精确的	
7	evolve /ɪ'vɒlv/	*v.* 发展；进化	
8	quench /kwentʃ/	*v.* 熄灭	
9	excessive /ɪk'sesɪv/	*adj.* 过多的，极度的	
10	connotation /ˌkɒnə'teɪʃn/	*n.* 内涵	
11	ritual /'rɪtʃuəl/	*n.* 仪式；惯例	
12	hospitality /ˌhɒspɪ'tæləti/	*n.* 好客；殷勤	
13	gratitude /'ɡrætɪtjuːd/	*n.* 感谢（的心情）；感激	
14	bride /braɪd/	*n.* 新娘	
15	groom /ɡruːm/	*n.* 马夫；新郎	
16	funeral /'fjuːnərəl/	*n.* 葬礼；麻烦事	
17	sacrifice /'sækrɪfaɪs/	*n.* 牺牲；祭品；供奉	
18	mourn /mɔːn/	*v.* 哀悼；忧伤；服丧	
19	exclusive /ɪk'skluːsɪv/	*adj.* 独有的；排外的	
20	majority /mə'dʒɒrəti/	*n.* 多数；成年	
21	individual /ˌɪndɪ'vɪdʒuəl/	*adj.* 个人的；个别的	
22	unadorned /ˌʌnə'dɔːnd/	*adj.* 朴素的；未装饰的	
23	worship /'wɜːʃɪp/	*n./v.* 崇拜；礼拜；尊敬	
24	Trading Routes of Silk Road	丝绸之路	
25	Tea Horse Road	茶马古道	
26	to trace back to	追溯到	
27	irrespective of	不管的，不考虑的	
28	Shennong	神农	
29	Daoist	道家	
30	Confucianist	儒家	

Task 2 | Comprehension

1 Note down the steps of making and drinking tea according to the passage.

_____

_____

_____

_____

_____

_____

_____

_____

2 Make a mind map of Chinese tea culture according to the passage above.

Chinese
Tea Culture

Section V
Insight into Proverbs

"Jie" 节—Integrity

"Jie" refers to reputation, integrity, strength, and moderation. The great spirit of self-abnegation, unceasing self-improvement, striving for success and the noble national integrity have become the spiritual mainstay of the nation self-strengthening and the pursuit of a person's personality.

1. 生存要素朴，气节要高尚。

Plain living and high thinking.

2. 不能正己，焉能化人。

Humanity means not judging or criticizing others.

When you can't even put yourself to rights, how can you hope to transform others?

(We should work on making ourselves better instead of trying to improve other people.)

3. 儒行之意味，简单地说，即尚气节、重力行。

The implication of "Confucian scholar practice" is simply to emphasize moral integrity and practice.

4. 孔子以 "仁" 为教育的核心，极为重视道德、诚信、气节、人格的教育。

Confucius, "benevolence" as the core of education, valued highly morality, honesty, integrity, and character education.

5. 崇高的民族气节是民族之魂，是民族尊严的体现，是民族生存和发展的根基。

The lofty moral courage is always considered as the soul of a nation and the base on which a nation exists and develops.

6. 师傅领进门，修行在个人。

Teachers open the door; you enter by yourself.

(The master leads the student through the door, but perfecting one's skill is up to the student.)

(A teacher can only expose students to knowledge; then it's up to the student to work hard to learn what he or she has been taught.)

7. 玉不琢，不成器。

If jade is not cut and polished, it can't be made into anything useful (and beautiful).

(You can't become anyone of consequence without the proper training and discipline.)

8. 有志不在年高，无志空活百岁。

Having aspirations has nothing to do with your age; without aspirations, you may live a hundred years for no purpose.

9. 将相本无种，男儿当自强。

Generals and prime ministers were not born to greatness; each man became so through great effort.

10. 人往高处走，水往低处流。

People need to ascend to a high place; water must flow ever downward.

(Said by parents to encourage their children to work to rise in life, rather than be like water that follows the natural law of gravity.)

11. 人过留名，雁过留声。

When a man passes away, all he leaves behind is his reputation;

when a wild goose passes by, all he leaves is his cry.

(A person's only legacy is his reputation.)

12. 名节如璧不可污。

A good name/reputation is not to be defiled.

13. 居安思危，戒奢以俭。

Stay vigilant to cope with troubles in times of safety; stay frugality to quit extravagance.

14. 富贵不能淫，贫贱不能移，威武不能屈。

Neither riches nor honours can corrupt him; neither poverty nor humbleness can make him swerve from principle; and neither threats nor forces can subdue him.

15. 饥不择食，寒不择衣，慌不择路，贫不择妻。

One who is starving is not picky about food.

(One who is freezing is not picky about clothing; one who is lost is not picky about the road he takes; one who is poor is not picky about a wife.)

(Beggars can't be choosers; when someone is in dire straits, he or she will often settle for a poor choice.)

Section VI
Self-assessment Checklist

1 Now, it's time for you to review your performance after learning this unit. Carry out a self-assessment by checking the following table.

Items	Ratings			
1. Knowledge	A	B	C	D
I know the development of Chinese food and drinking culture.				
I know the features of regional cuisines and the art of Chinese tea making.				
I master the basic factors of regional cuisines and customs of drinking tea.				
I know the etiquette of table manners and drinking Chinese tea.				
I master the expressions of Chinese dishes.				
2. Skills	A	B	C	D
I can use appositive to help understand.				
I can identify non-restrictive attributive clauses.				
I master the function of non-predicate verb "ving" as a post-positive attribute.				
I can search information by using skimming and scanning.				
3. Speaking	A	B	C	D
I can talk one of the Chinese dishes and its features fluently.				
I can explain the reasons of some Chinese sayings about food logically.				
I can critically analyze and tell food problems posted in Internet.				
I can tell the procedure of making a dish.				
4. Confidence in Chinese Culture	A	B	C	D
I can understand the essence of Chinese food and tea art.				
I appreciate Chinese tea drinking.				
I am proud of Chinese food culture.				

A: Basically agree

B: Agree

C: Strongly agree

D: Disagree

2 Write down the corresponding English dishes for the Chinese menu below.

老街大排档

招牌菜

老街秘制鱼 48元（送花甲、馍干）单点配菜（5元/份）

（鸭血、豆腐、豆腐皮、青菜、土豆、莲菜）

老街松滋鸡 58元　　　　　　　老街一品香 38元

荤菜		素菜	
干煸香酥虾	48元/份	烧时蔬	12元/份
辣炒花甲	28元/份	烧腐竹	16元/份
小炒羊肉	46元/份	韭菜鸭血	16元/份
肉沫酸豆角	22元/份	豆芽面筋	12元/份
萝卜干炒腊肉	28元/份	青菜炕豆腐	16元/份
花肉老北瓜	26元/份	干炸蘑菇	16元/份
水煮肉片	28元/份	老街地三鲜	16元/份
老街回锅肉	26元/份	番茄鸡蛋	12元/份
糖醋里脊	28元/份	豆角炒茄子	16元/份
鱼香肉丝	26元/份	干锅土豆片	16元/份
老街木须肉	28元/份	酸辣土豆丝	12元/份
酸菜烧牛肉	46元/份	鱼香茄子（红烧）	16元/份
红烧大肠（干煸）	36元/份	萝卜炒粉条	12元/份
孜然鱿鱼	46元/份	鸡蛋粉条	12元/份
宫保鸡丁	26元/份	时蔬烧蘑菇	16元/份

汤类

酸辣肚丝汤 26元/份　　　　一盆豆腐汤 16元/份

百合绿豆汤 16元/份　　　　西湖牛肉羹 28元/份

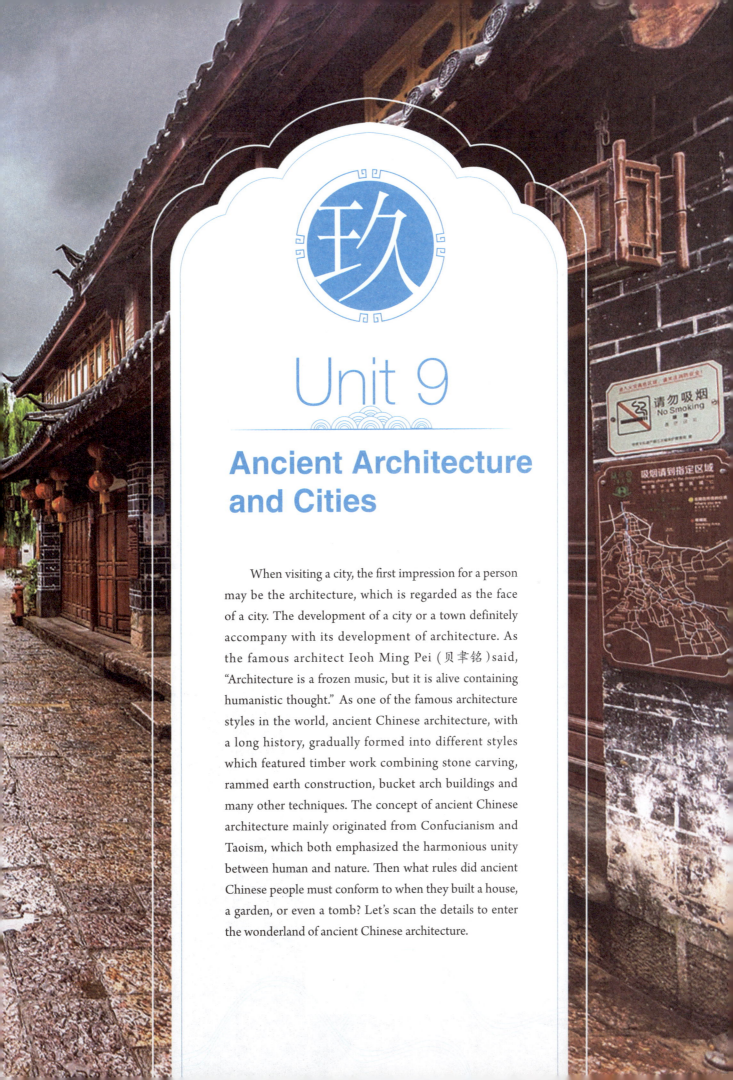

玖

Unit 9

Ancient Architecture and Cities

When visiting a city, the first impression for a person may be the architecture, which is regarded as the face of a city. The development of a city or a town definitely accompany with its development of architecture. As the famous architect Ieoh Ming Pei (贝聿铭) said, "Architecture is a frozen music, but it is alive containing humanistic thought." As one of the famous architecture styles in the world, ancient Chinese architecture, with a long history, gradually formed into different styles which featured timber work combining stone carving, rammed earth construction, bucket arch buildings and many other techniques. The concept of ancient Chinese architecture mainly originated from Confucianism and Taoism, which both emphasized the harmonious unity between human and nature. Then what rules did ancient Chinese people must conform to when they built a house, a garden, or even a tomb? Let's scan the details to enter the wonderland of ancient Chinese architecture.

Overview

Ancient Architechture and Cities

Introduction
- Chinese Architecture
- Learning Objectives

Brainstorming
- Identify Pictures of Buildings
- Know the Culture of Tian Tan
- Critical Thinking

Story Reading
- Text Reading: Chinese Architecture
- Vocabulary
- Expressions for Etiquette of Chinese Architecture
- Consolidated Exercises

Let's Practice
- Know to Talk About a City
- Team Work

Extended Reading
- Text: The Hanging Monastery
- Comprehension
- Picture Drawing

Insight into Proverbs

Self-assessment Checklist
- Etiquette Observation
- Summary

Learning Objectives

After learning this unit, students are able to reach the goals below.

1	专业能力目标	① 了解中国古代建筑的哲学理念及其对人们生活的影响
		② 扩大与中国古代建筑相关的词汇量，学会相关句型的表达
		③ 提升阅读速度，能较流利地讲述中国古代建筑的分类及其特征
2	方法能力目标	① 综合运用阅读技巧，掌握中国古代建筑所体现的思想
		② 运用语法功能解决阅读问题
		③ 掌握下定义或概念在阅读理解上的功能
		④ 综合运用构词法、联系上下文法、同义词或反义词法，以此来扩大词汇量
3	社会能力目标	① 通过主题阅读，形成丰富的空间想象力，铸就创新能力
		② 通过网络获取不同风格建筑的信息，能辨别和分析不同风格的建筑
		③ 掌握英语阅读的技巧，提高学习效率和社会工作效率
		④ 通过主题阅读，能运用恰当的基本知识和语言讲述中国经典建筑特色及其蕴含的国人智慧，能欣赏不同建筑蕴含的文化，并能恰当地运用这些国粹提升自身阅历
4	情感与思政目标	① 通过本单元的学习，加深对中国经典建筑及其折射出的经典中国文化的认知；加强爱护、维护公物和建筑古迹的意识
		② 通过主题阅读，懂得中国传统的建筑智慧；开拓文化视野，增强文化自信，培养建筑审美情趣，提高人文理念，正确树立"崇尚劳动、尊重劳动"的观念；树立劳动光荣的价值观
		③ 通过主题阅读及拓展阅读，加强运用中国建筑哲理分析、探究人类发展生存与自然生态之间问题的认识；具备团结合作的精神和创新能力

Section I
Brainstorming

Task 1 | **Let's Warm Up**

Look at the pictures below and match them with the correct choices provided.

(1) The Great Wall　　(2) Stilted Building　　(3) Huizhou Building　　(4) The Forbidden City

(5) Tulou　　(6) Pagoda　　(7) Wind and Rain Bridge of Dong Ethnic Group

(8) Suzhou Garden　　(9) Memorial Arch

1   Watch the video and fill in the blanks with proper words.

Emperors held annual ceremonies to pray to heaven for a good harvest at the Temple of Heaven, Tiantan. It was _____(1)_____ in Beijing in 1420. Emperors prayed to heaven for good harvests. Ancient Chinese believed that the earth was _____(2)_____ and heaven was _____(3)_____. The altars and palaces were built into circles to represent _____(4)_____. The southern wall was rectangular to represent the _____(5)_____. Chinese show their _____(6)_____ to heaven in Tiantan. The blue-glazed tiles of the imperial vault of heaven represent heaven. The circular mound altar is surrounded by a ring of _____(7)_____ plates, which are also surrounded by multiple nine plates. It represents the supremacy of _____(8)_____. Tiantan covers an area of 2.7 million square of meters with an orderly arrangement showing Chinese regality.

2   Answer the questions after watching the video.

(1) What did ancient Chinese people think of the heaven and the earth?

(2) What did emperors do in Tiantan?

(3) What is the shape of the construction of Tiantan?

3   Critical thinking.

(1) Why was Tiantan built in the shape of round?

(2) What does number "nine" represent in Chinese culture?

Section II
Story Reading

Task 1 **Let's Read**

Ancient Chinese Architecture

Ancient Chinese architecture has a long history which can be traced back to the Shang Dynasty (1600 BC—1046 BC). Along the history, Chinese ancient architecture has formed rich and varied styles, such as temples, imperial palaces, altars, pavilions, official residences and folk houses, which greatly reflect ancient thought—the harmonious unity of human beings with nature. It is generally grouped into imperial architecture, religious architecture, garden architecture and general architecture.

Chinese imperial architecture includes imperial palaces, gardens and mausoleums. To incarnate the supremacy of imperial power, it often adopts the layout of an axial symmetry, with the buildings on the central axis tall and splendid while the rest rather small and simple. The Forbidden City is an example.

Traditional Chinese Residences refer to those places for common people to live in. As a multiple-ethnic country, Chinese people have weaved a colorful civilian residence picture. From the distinctive houses in traditional courtyards (siheyuan) in Beijing, farmers' caves (yaodong) in Shaanxi Province, stilted building (diaojiaolou) on steep inclines or projecting over water in southern China, seal-like compound (yikeyin) in Yunnan Province, to the earthen buildings (tulou) of Hakkas in Fujian Province, ancient Chinese people showed their wisdom in creating wondrous architectural styles based on their geographical features, climates, customs, and habits.

Traditional Gardens are famous for their variety and delicate craftwork. Generally they can be divided into two groups, the Imperial Garden Architecture in north China and the Private Garden Architecture in south China. Imperial gardens are noteworthy for their grand dimensions, luxurious buildings, and exquisite decorations, like the Summer Palace and Old Summer Palace (Ruins of Yuanmingyuan). While the gardens in the south are smaller but no less exquisite, like a shy girl waiting for you to take away her veiling. Most of the southern gardens are scattered in Jiangsu and Zhejiang provinces. Lingering Garden and Humble Administrator's Garden are the noteworthy.

Religious architecture in China has an added flavor to it. Different religions have their unique architectural styles to represent their different ideals.

Whatever the architecture is, their most significant characteristic is the use of timber framework. Paintings and carvings are added to the architectural work to make it more beautiful and attractive. It has its own principles of structure and layout.

Ancient Chinese architecture embodies Chinese ritual system thoughts. It stresses the hierarchical idea, pattern, scale, color, structure; all of these have strict regulations. Besides, the ideology of the unity of heaven and man is also demonstrated in the development process of ancient Chinese architecture, which promotes the mutual coordination and integration of architecture and nature.

Another principle is fengshui. Architectural fengshui is the soul of China's ancient architectural theories. It enjoys a high status in China's architecture history. Fengshui is literally translated as Wind Water. It is a method of investigating geographical features used in site selection and construction of dwellings, cities, and tombs, etc. by means of examining soil and tasting water. Its core concept is also the harmonious coexistence between human beings and nature. (513 words)

Task 2 Let's Learn

1 Read the following words and expressions, and tick the ones you know in the last column of the word list.

Numbers	Words & Expressions	Meanings	Notes
1	architecture /ˈɑːkɪtektʃə(r)/	n. 建筑学；建筑风格	
2	altar /ˈɔːltə(r)/	n. 祭坛；圣坛	
3	pavilion /pəˈvɪliən/	n. 阁；亭子	
4	residence /ˈrezɪdəns/	n. 住宅，住处	
5	mausoleum /ˌmɔːsəˈliːəm/	n. 陵墓	
6	incarnate /ɪnˈkɑːnət/	v. 体现，化身为	

Numbers	Words & Expressions	Meanings	Notes
7	supremacy /suːˈpreməsi; sjuːˈpreməsi/	n. 霸权；至高无上；主权	
8	adopt /əˈdɒpt/	v. 采取；接受	
9	layout /ˈleɪaʊt/	n. 布局；设计	
10	axial /ˈæksɪəl/	adj. 轴的；轴向的	
11	symmetry /ˈsɪmətri/	n. 对称（性）；整齐	
12	multiple-ethnic /ˈmʌltɪpl-ˈeθnɪk/	adj. 多民族的	
13	civilian /səˈvɪliən/	n. 平民，百姓	
14	distinctive /dɪˈstɪŋktɪv/	adj. 独特的，有特色的；与众不同的	
15	wondrous /ˈwʌndrəs/	adj. 奇妙的；令人惊奇的；非常的	
16	delicate /ˈdelɪkət/	adj. 微妙的；精美的，雅致的	
19	craftwork /ˈkrɑːftwɜːk/	n. 工艺，工艺品	
20	Hakka /ˈhækə/	n.（汉）客家；客家人；客家语	
21	noteworthy /ˈnəʊtwɜːði/	adj. 值得注意的；显著的	
22	dimension /daɪˈmenʃn; dɪˈmenʃn/	n. 方面；[数] 维；尺寸	
23	exquisite /ɪkˈskwɪzɪt; ˈekskwɪzɪt/	adj. 精致的；细腻的；优美的	
24	luxurious /lʌɡˈʒʊəriəs/	adj. 奢侈的；丰富的	
25	decoration /ˌdekəˈreɪʃn/	n. 装饰，装潢；装饰品	
26	veiling /ˈveɪlɪŋ/	n. 面纱；面纱布料	
27	scatter /ˈskætə(r)/	v. 撒播；散开；散布	
28	unique /juˈniːk/	adj. 独特的，稀罕的	
29	significant /sɪɡˈnɪfɪkənt/	adj. 重大的；有效的；有意义的	
30	timber /ˈtɪmbə(r)/	n. 木材；木料	
31	framework /ˈfreɪmwɜːk/	n. 框架，骨架	
32	carve /kɑːv/	v. 雕刻；切开	
33	ritual /ˈrɪtʃuəl/	n. 仪式；惯例；礼制	
34	hierarchical /ˌhaɪəˈrɑːkɪkl/	adj. 分层的；等级体系的	
35	regulation /ˌreɡjuˈleɪʃn/	n. 管理；规则；校准	
36	promote /prəˈməʊt/	v. 促进；提升	
37	mutual /ˈmjuːtʃuəl/	adj. 共同的；相互的	
38	coordination /kəʊˌɔːdɪˈneɪʃn/	n. 协调，调和	
39	integration /ˌɪntɪˈɡreɪʃn/	n. 集成；综合	
40	dwelling /ˈdwelɪŋ/	n. 住处；寓所	

2 Read and learn the key terms of traditional Chinese architecture, and tick the ones you know in the last column of the word list.

Numbers	Terms	Meanings	Notes
1	the harmonious unity of human beings with nature	人与自然的和谐统一	
2	The Forbidden City	故宫（紫禁城）	
3	courtyards (siheyuan)	四合院	
4	farmers' caves (yaodong)	窑洞	
5	stilted building (diaojiaolou)	吊脚楼	
6	seal-like compound (yikeyin)	一颗印	
7	the earthen buildings (tulou)	土楼	
8	the Summer Palace	颐和园	
9	the Old Summer Palace (Ruins of Yuanmingyuan)	圆明园	
10	fengshui	风水	

3 Get to know some useful expressions for etiquette of Chinese architecture.

Some etiquette cultures are very common in the layout of ancient Chinese architecture, which has been inherited for thousand years. It shows respect for the elders and religious ancestors. It's a reflection of Chinese ritual culture. The rhythm, harmony, contrast, symmetry, axis and other design techniques have been used in ancient buildings in China to achieve the best viewing effect and maintain reasonable functions.

(1) Confucianism and Taoism etiquette principles

① 尊崇主体、旁侧从属（长者为先、主从有序）

To adopt the layout of an axial symmetry, with the buildings on the central axis tall and splendid while the rest rather small and simple.

② 人与自然的和谐统一

To emphasize the harmonious unity of human beings with nature.

(2) Fengshui principles

依山傍水　leaning against mountains and facing waters

坐北朝南　facing south

因地制宜　in accordance with local conditions

两边对称　bilaterally symmetrical

曲径通幽　a winding path leading to a secluded spot

Task 3　Consolidated Exercises

1　Answer the following questions briefly according to the passage.

(1) How many types of house are usually divided into according to the passage?

(2) What are the characteristics of imperial architecture?

(3) What are the reasons for traditional residences colorful and various?

(4) What are the features of gardens in southern China?

(5) What are the common features of ancient architecture?

(6) What ritual system thoughts did ancient architecture follow?

(7) What do you think of the ritual system thoughts that ancient architecture followed?

(8) How do you think of the principles of fengshui work in ancient architecture?

2　Fill in the blanks with the proper words given, and change the word form if necessary.

delicate	adopt	promote
residence	embody	dwelling
wondrous	mutual	

(1) They have been denied ＿＿＿＿＿＿ in this country.

(2) The state will ＿＿＿＿＿＿ a more restrictive policy on arms sales.

(3) Oh, it was ＿＿＿＿＿＿ beautiful!

(4) Tourists often disturb the ＿＿＿＿＿＿ balance of nature on the island.

(5) The meeting discussed how to ＿＿＿＿＿＿ this latest product.

(6) All you have to do is to unite mentally and emotionally with the good you wish to ＿＿＿＿＿＿.

(7) Our relationship was based on ＿＿＿＿＿＿ dependence.

(8) After ＿＿＿＿＿＿ in Rio for ten years, she moved to Lisbon.

3　Translate the following paragraph into English.

苏州是中国著名的"园林城市 (city of gardens)"。其园林艺术已有1,500年的历史。14世纪到20世纪之间的明清时期是其园林建筑的黄金时期。这座城市里曾经有超过200家私人园林。这些精致小巧的园林吸引了大量游客。1997年苏州园林被列入世界文化遗产名录 (the List of World Cultural Heritage)。

Let's Practice

Task 1 Let's Talk

1 Watch and learn.

Watch the video to know how to talk about a city/town/mountain/lake/river and learn some expressions.

(1) Location:

be located in…; lie in the east/south of…; be situated in…

(2) History or original story:

Legend says that...

It has a history of… years.

(3) Attractions:

The Leifeng Tower is a popular attraction at…

The most attraction in… is…

(4) Characteristics:

The picturesque is surrounded by elegant landscapes, such as…

(5) Evaluation:

As a famous resort, … is a perfect combination of artificiality and nature.

It looks like a Chinese traditional painting.

2 Imitate the above video to introduce your hometown using the expressions correctly and fluently.

Task 2 Team Work

1 Presentation.

Choose one style of ancient Chinese architecture to talk in groups and then present it in class.

2 Place introduction.

Work in groups to talk about a place you've ever been to or you are eager to visit.

3 Group discussion.

Watch a video of fengshui and then discuss "Some people think that fengshui is superstition. Do you think so? How did fengshui work in ancient Chinese architecture?"

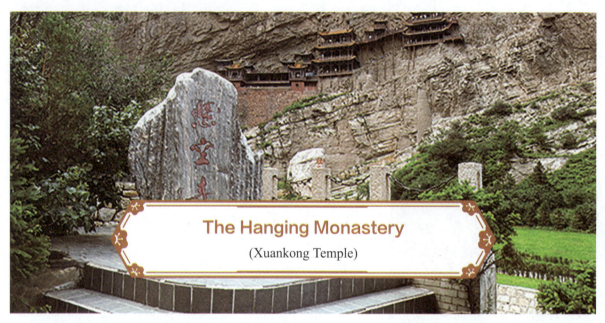

The Hanging Monastery
(Xuankong Temple)

In the Chinese creation myth, Pan Gu's head and limbs incarnated into the Five Great Mountains after he departed, namely East Mount Tai, West Mount Hua, South Mount Heng, North Mount Heng, and Middle Mount Song. They are divine, magnificent, and spectacular, where they were believed as the path to heaven or the residence of Deities. Therefore, in those great mountains, emperors held grand ceremonies to worship heaven or pray for blessing, and Taoists and Buddhists built temples to practice.

Mount Heng, located in Shanxi Province, north of China, is famous for a unique temple. The temple, hanging on the cliff of Mount Heng, is called the Hanging Monastery (Xuankong Temple), which was built in the later period of Northern Wei Dynasty (368—534). It is a rare piece of architecture of its kind in China. It was listed as one of the 10 top architectural wonders of the world by the *Time* magazine in Dec. 2010.

The Hanging Monastery is an architectural wonder. Suspended some 50 meters over the ground and consisting of 40 pavilions and halls, the Hanging Monastery is constructed with the help of wooden pillars that are anchored into the cliff face behind the mountain. A unique mechanical theory was applied to building the framework. Crossbeams were half-inserted into the rock as the foundation, while the rock in back became its support. These pavilions and halls are intricately and ingeniously linked by winding corridors, bridges, and boardwalks that offer a perilous glimpse of the ground below, well interpreting a local saying "hung by three horsetails suspending in the air". Seen from below, it appears to be a tumble-down castle in the air. Inside, it provides the same scene as other temples.

There are more than 80 cast bronze statues, cast iron statues, and clay sculptures and stone carvings handed down from different dynasties inside the temple. The most outstanding feature of the Hanging Monastery is the sculptures of Sakyamuni, founder of Buddhism, Confucius, founder of Confucianism, and Lao Tzu, the Taoism master, being presented side-by-side in the Three Saints Hall, which is unusual.

The Hanging Monastery is considered as the best place to explore the facts of Chinese religions. The coexistence of different religions embodies the Confucian thought "propriety first", "benevolence", "wisdom" and the Taoist "infinite", "harmony" and the thought of Buddhism "universal salvation", setting an example for guiding people to resolve conflicts and disputes.

Same as other mountains, Mount Heng is a place full of beautiful legends. In Chinese mythology, it is said that many well-established people completed their practices and became immortals there.

Being fantastic, dangling and ingenious in construction and multi-religions, the Hanging Monastery won its world fame. Some construction experts from countries including Britain, Germany, and Italy said that the Hanging Monastery in China combines mechanics, aesthetics and religion into one, achieving perfection, which is rarely seen in the world. It is not only the pride of the Chinese people, but also the pride of the people of the world. (531 words)

Task 1　New Words and Expressions

Read the following words and expressions, and tick the ones you know in the last column of the word list.

Numbers	Words & Expressions	Meanings	Notes
1	limb /lɪm/	n. 肢，臂	
2	magnificent /mægˈnɪfɪsnt/	adj. 高尚的；壮丽的	
3	spectacular /spekˈtækjələ(r)/	adj. 壮观的，惊人的	
4	deity /ˈdeɪəti; ˈdiːəti/	n. 神；神性	
5	worship /ˈwɜːʃɪp/	n./v. 崇拜；礼拜	
6	pray /preɪ/	v. 祈祷；恳求	
7	bless /bles/	vt. 祝福；保佑	
8	cliff /klɪf/	n. 悬崖；绝壁	
9	suspend /səˈspend/	v. 延缓，推迟；暂停；悬浮	
10	pavilion /pəˈvɪliən/	n. 阁；亭子	
11	anchor /ˈæŋkə(r)/	n. 锚	
12	intricately /ˈɪntrɪkətli/	adv. 杂乱地	
13	ingeniously /ɪnˈdʒiːniəsli/	adv. 贤明地；巧妙地	
14	sculpture /ˈskʌlptʃə(r)/	n. 雕塑；雕刻	
15	exclusively /ɪkˈskluːsɪvli/	adv. 唯一地，排外地	
16	invader /ɪnˈveɪdə(r)/	n. 侵略者；侵入物	
17	pilgrimage /ˈpɪlɡrɪmɪdʒ/	n. 朝圣　vi. 朝拜	
18	wane /weɪn/	vi. 衰落；变小	
19	resolve /rɪˈzɒlv/	vt. 决定；溶解	
20	dispute /dɪˈspjuːt; ˈdɪspjuːt/	n./v. 辩论；争吵	
21	mythology /mɪˈθɒlədʒi/	n. 神话；神话学	
22	complete /kəmˈpliːt/	adj. 完整的；完全的	
23	immortal /ɪˈmɔːtl/	adj. 不朽的；神仙的；长生的	
24	dangling /ˈdæŋɡ(ə)lɪŋ; ˈdæŋɡlɪŋ/	adj. 悬挂的；摇摆的	
25	ingenious /ɪnˈdʒiːniəs/	adj. 有独创性的；机灵的，精制的	
26	aesthetic /esˈθetɪk/	adj. 美的；美学的	
27	creation myth	创世神话	
28	Pan Gu	盘古	
29	Sakyamuni	释迦牟尼	
30	the Five Great Mountains	五岳	
31	the Three Saints Hall	三圣殿	
32	The Hanging Monastery	悬空寺	
33	Universal Salvation	普度众生	

Task 2 Comprehension

1 Answer the questions according to the passage.

(1) How did the Five Great Mountains come into being according to the legend ?

(2) What did the emperors usually do in the Great Five Mountains?

(3) Why can the Hanging Monastery be listed as one of the 10 top architectural wonders of the world?

(4) What is the construction of the temple?

(5) How many pavilions and sculptures are there in the temple?

(6) What is the most unique feature in the Three Saints Hall?

(7) Who are Confucius, Lao Tzu, and Sakyamuni?

(8) How did some foreign construction experts think of the Hanging Monastery?

2 Draw a picture of the construction of the Hanging Monastery according to the passage.

儒家的伦理范畴把智、

勇敢。孔子认为

苟施「仁」的条件之

「勇」

语·阳货》……

勇而无义为乱。」又论

语·子罕》：：「知者不惑，

仁者不忧，勇者不惧。」

Section V
Insight into Proverbs

"Yong"勇—Bravery

"Yong" refers to bravery, strength, and fortitude. "Yong" means daring to think, to break new ground, to innovate, to exploit, to assume responsibility, to stick to principles, and to seek truth from facts. The unyielding, heroic and indomitable spirit embodied in traditional Chinese culture is the inexhaustible driving force for the Chinese people, who have gone through untold hardships.

1. 好汉做事好汉当。

A great person has the courage to accept the consequences of his own actions.

2. 吃得苦中苦，方为人上人。

Only if you have endured the bitterest suffering can you become a superior person.

3. 初生牛犊不怕虎。

A newborn calf has no fear of tigers./The more wit, the less courage.

(Youth is fearless.)

4. 逆水行舟，不进则退。

Learning is like rowing a boat against the current; if you don't advance, you'll regress.

5. 愚公移山。

The foolish old man moved the mountain.

(Anything can be done if you work long and hard enough at it.)

6. 只要功夫深，铁杵磨成针。

With enough work, an iron rod can be ground into a needle.

(Almost anything can be achieved if you put enough effort into it.)

7. 好事多磨。

Good things are produced only through much grinding.

(Nothing good can be accomplished without a lot of work and many setbacks.)

8. 失败是成功之母。

Failure is the mother of success.

9. 万事开头难。

Everything is hard in the beginning.

10. 好勇疾贫，乱也。

A man of courage who hates to be poor will be sure to commit a crime.

11. 好汉不提当年勇。

A great man is silent about his past glories.

(Truly great people never dwell on their past glories.)

12. 不入虎穴，焉得虎子？

If you don't enter the tiger's den, how will you ever get the tiger's cubs?

(Nothing ventured, nothing gained; No pain, no gain.)

13. 后生可畏，勇者不惧。

Youths should be respected; men of courage are free from fear.

14. 三军可夺帅也，匹夫不可夺志也。

The general of an army may be carried off, but a man of the

common people cannot be robbed of his free will.

15. 勇敢面前没有通不过的路。

No way is impossible to courage.

Section VI
Self-assessment Checklist

1 Now, it's time for you to review your performance after learning this unit. Carry out a self-assessment by checking the following table.

Items	Ratings			
1. Knowledge	A	B	C	D
I know the styles of ancient Chinese architecture.				
I know the features of ancient Chinese architecture and the Hanging Monastery.				
I master the ancient Chinese architecture's connotation culture.				
I know the principles that ancient Chinese architecture conforms to.				
I master the expressions of Chinese architecture.				
2. Skills	A	B	C	D
I can use grammar functions to solve reading problems.				
I can use word formation, context, synonym and antonym in the reading.				
I grasp the functions of definitions or concepts in reading.				
3. Speaking	A	B	C	D
I can talk one type of Chinese architecture fluently.				
I can explain the culture of Chinese architecture logically.				
I can think critically and tell about fengshui in Chinese architecture.				
I can tell the principles of Confucianism and Taoism in Chinese architecture.				
4. Confidence in Chinese Culture	A	B	C	D
I have the awareness of protecting Chinese architecture art.				
I have the aesthetic and appreciation of Chinese architecture art.				
I can integrate traditional Chinese architecture with the western one or modern one.				

A: Basically agree

B: Agree

C: Strongly agree

D: Disagree

2 Fill in the blanks in the mind map below to check whether you have a good understanding of this unit.

(1) Draw a mind map of ancient Chinese architecture, including its styles, features, and distributions.

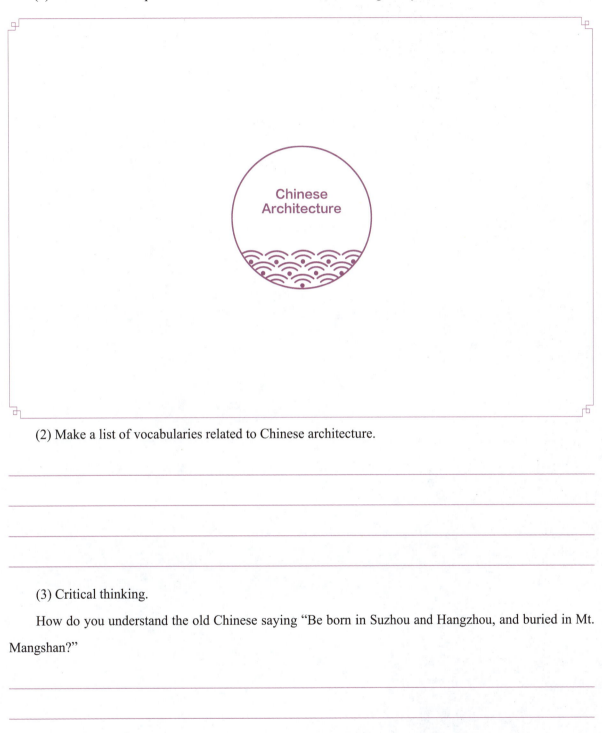

(2) Make a list of vocabularies related to Chinese architecture.

_____

_____

_____

_____

(3) Critical thinking.

How do you understand the old Chinese saying "Be born in Suzhou and Hangzhou, and buried in Mt. Mangshan?"

_____

_____

_____

_____

Unit 10

Chinese Ancient Entertainments

Chinese leisure activities have accompanied people's life from the early history. According to the book *Strategies of the Warring States* (《战国策》), it is recorded that Linzi was very rich. The people in Linzi played lutes, blew pipe instruments, bet cockfighting, trained dogs and played Cuju. Just as Kiyoshi Mik (三木清) ever said "Entertainment exists in life and creates the style of life." Chinese ancient entertainments usually originated from labour and daily life. Ancient Chinese people entertained themselves by playing a drinkers' wager game during a banquet, playing Chinese Go when paying a visit, or even betting a cricket/cock fighting in the market. What games were popular in ancient China? Do you like playing football? Do you know what football was called in China in the past? In traditional Chinese culture, do you know how Chinese play for relaxation and fun? In this unit, the traditional Chinese entertainments are introduced. Let's scan the overview to know the details.

Overview

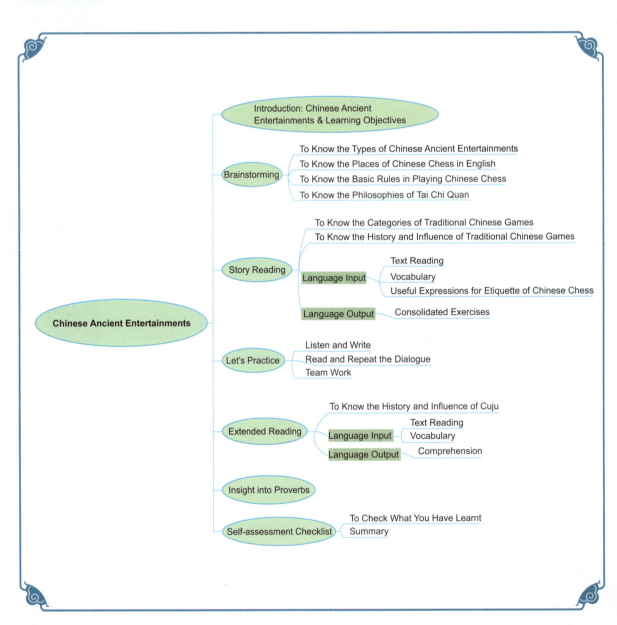

Chinese Ancient Entertainments

- Introduction: Chinese Ancient Entertainments & Learning Objectives

- Brainstorming
  - To Know the Types of Chinese Ancient Entertainments
  - To Know the Places of Chinese Chess in English
  - To Know the Basic Rules in Playing Chinese Chess
  - To Know the Philosophies of Tai Chi Quan

- Story Reading
  - To Know the Categories of Traditional Chinese Games
  - To Know the History and Influence of Traditional Chinese Games
  - Language Input
    - Text Reading
    - Vocabulary
    - Useful Expressions for Etiquette of Chinese Chess
  - Language Output
    - Consolidated Exercises

- Let's Practice
  - Listen and Write
  - Read and Repeat the Dialogue
  - Team Work

- Extended Reading
  - To Know the History and Influence of Cuju
  - Language Input
    - Text Reading
    - Vocabulary
  - Language Output
    - Comprehension

- Insight into Proverbs

- Self-assessment Checklist
  - To Check What You Have Learnt
  - Summary

Learning Objectives

After learning this unit, students are able to reach the goals below.

1	专业能力目标	① 通过对中国象棋、围棋、毽子、射箭、太极拳和蹴鞠的学习，了解中国古代娱乐活动的起源、发展及其对人们生活的影响和作用
		② 通过对象棋棋子和下棋规则等的了解，掌握与中国古代棋类活动相关的词汇、句型表达
		③ 通过快速阅读法和信息捕捉法，掌握文章主旨和大意，理解中国古代娱乐活动的发展历史；通过抓取文章的核心思想、行文思路、语言技巧及文体修辞等，进一步提高阅读理解能力和思想表达能力
2	方法能力目标	① 运用总分法阅读文章，抓取文章的主旨和大意，理解围棋、象棋、太极拳、毽子和蹴鞠等娱乐活动的作用；掌握阅读材料所体现的思想观点，体悟"乐以养性"之道
		② 养成逻辑推理的好习惯，能根据中国古代娱乐专题阅读，总结中国古代娱乐活动中的运转规则，并付诸实践
		③ 在规定时间内，能运用略读、寻读、推测等方法最大化地摄取核心信息，了解要点与观点
		④ 学会在阅读中通过构词法、行文逻辑、上下文语境等方法猜测单词意义，扩大词汇量
3	社会能力目标	① 通过太极之道、下棋之规等，锻炼发散思维与逻辑批判思维，提高想象力与创新能力
		② 学会遵守社会规则，在纷繁复杂的社会环境中，具备遇事冷静思考的能力
		③ 掌握中国古代娱乐活动中的哲学原理，提高学习和工作效率
		④ 通过文章阅读，扩大知识面，从而能运用恰当的基本知识和语言讲述中国古代娱乐活动的哲学逻辑和国人智慧；能恰当地运用这些娱乐活动提升自己的生活质量
4	情感与思政目标	① 通过本单元的学习，掌握中国古代娱乐活动中的规则，理解中国古代娱乐活动对生活和社会的重要意义，加强娱乐文化话语模式与娱乐化思维方式的判断能力、思考能力
		② 通过主题阅读，形成正确的娱乐观、道德人格与道德行为；能做到遵规守纪；具有正义文明的道德情操
		③ 通过主题及拓展阅读，能协调娱乐与学习、工作之间的关系；能明辨并选择健康的娱乐方式；形成正确的世界观、人生观与价值观

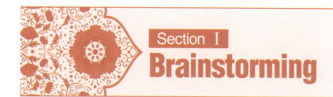

Section I
Brainstorming

Task 1 **Let's Warm Up**

1 Look at the pictures below and describe the things in English in the pictures.

2 Write down the basic rules for the pieces in Chinese Chess.

Pieces	Basic Rules
帅 将	
士 士	
馬 馬	
炮 炮	
相 象	
車 車	
兵 卒	

Task 2　Let's Watch and Say

1 Watch the video Tai Chi and try to write down the missing words in the blanks.

The Yin-yang philosophy of Tai Chi includes static (静), dynamic (动), fast and _____(1)_____. Chinese believe that dynamic is contained in static and fast is contained in slow.

Tai Chi is the combination of these _____(2)_____. It's fighting principle is to maneuver (调动) the internal energy flows to achieve _____(3)_____ and power. The most important function of Tai Chi is to _____(4)_____ the theory of static and dynamic in nature and life and tune (调整) one's heart and figure in _____(5)_____ with nature.

2 Critical thinking.

(1) Do you know what philosophies are contained in Chinese Chess and Chinese Go (Weiqi)?

(2) How do you understand "retracting a false move in a chess game"? Do you allow that to happen in playing chess? Why?

_____

_____

_____

_____

_____

_____

_____

_____

_____

_____

_____

Section II
Story Reading

Task 1 Let's Read

Traditional Chinese Games

Chinese games are a very important part of Chinese culture. When strolling through the streets in China, you can see and hear people playing Chinese traditional games for relaxation, fun and more. You can also see plenty of interested onlookers watching and commenting on the strategies and moves of the players. A brief introduction tells what games Chinese play for relaxation and mental stimulation.

Ancient Chinese games like chess (Xiangqi), encirclement chess (Weiqi), which is similar to chess, and checkers (Tiaoqi) are popular board games. They are not only enjoyable, but they're challenging. They therefore require brain usage, which is why middle-aged and elderly Chinese in particular like to play them regularly. According to traditional Chinese longevity principles, playing games that are fun, relaxing and mentally challenging is an ideal way to improve and maintain your mental health and to help slow the decline of old age.

Chinese Chess

Xiangqi, or Chinese Chess, is an extremely popular game in the world, especially in the eastern hemisphere. It is currently played by millions (or tens of millions) in China, Thailand, Singapore, Vietnam and other Asian countries. Xiangqi has remained in its present form for centuries.

In China, you often can see people play the game on street using a large chess set. Some players may hit the board hard at some moves to show his power. Since the pieces are large (about 2 inches in diameter), the effect is dramatic. They may also say or sing something to do the trick. Usually there are a few people watching the game. If they know both players (sometimes even they don't know the players), they may point out some moves for one or both players. To prevent the helpers to say anything during a game, the players often remind them by saying "watching but not telling, a true gentleman".

Chinese Go (Weiqi)

Weiqi or Go is most popular in China, Japan, and Korea. Go is a much deeper game than Chinese Chess. Chinese Chess is more a folk game and Go is more popular in those with higher education. While observing, you can guess the next move the player might move, so you can learn from it. The thing is the game really get your attention and you have to think about it. The game is just like life. We start from

nothing, then we add a little bit more at each move. We make some good moves and some bad moves. At one point, it becomes so complicated that we can't even figure out what to do or what's the best move, but we have to. We often call Go "the black and white world".

Shuttlecock Kicking

Shuttlecock (Jianzi) is a traditional Asian game in which players aim to keep a heavily weighted shuttlecock in the air by using their bodies, apart from the hands. The game is played on a court similar to badminton, or played artistically, among a circle of players in a street or park. The first known version of Jianzi was in the 5th century BC in China. Over the next 1000 years, this shuttlecock game spread

throughout Asia. Thus the game has a history of more than two thousand years. Kicking shuttlecock is vigorous exercise, and it helps to build hand-eye coordination which is good for the health of the brain.

Archery

China is one of the first countries in the world to have and use the bow and arrow. In the Western Zhou Dynasty, archery was considered a basic skill in the aristocracy and was a very important educational subject. The Chinese philosopher Confucius actively promoted and wrote about the use of the bow and

arrow. Archery is one of the six skills in Confucius teachings. If there were some disputes, it must be settled with an archery competition. Confucius paid much attention to the educational function of archery, which was a useful technique to learn etiquette and to restrain people's behaviours. (662 words)

Task 2 　Let's Learn

1 Read the following words and expressions, and tick the ones you know in the last column of the word list.

Numbers	Words & Expressions	Meanings	Notes
1	stroll /strəʊl/	v./n. 闲逛；溜达	
2	relaxation /ˌriːlækˈseɪʃn/	n. 放松；休息；消遣	
3	onlooker /ˈɒnlʊkə(r)/	n. 旁观者	
4	comment /ˈkɒment/	n. 议论；评论；解释	
5	mental /ˈmentl/	adj. 思想的；精神的	
6	stimulation /ˌstɪmjʊˈleɪʃən/	n. 激励；兴奋	
7	encirclement /ɪnˈsɜːklmənt/	n. 合围；包围	
12	longevity /lɒnˈdʒevəti/	n. 长命；长寿；持久	
13	principle /ˈprɪnsəpl/	n. 原则；规范；信条	
14	ideal /aɪˈdiːəl/	adj. 完美的；理想的	
15	decline /dɪˈklaɪn/	n./v. 衰退；衰落	
16	currently /ˈkʌrəntli/	adv. 现在；目前；当前；当下	
17	diameter /daɪˈæmɪtə(r)/	n. 直径	
18	dramatic /drəˈmætɪk/	adj. 突然的；令人吃惊的	
19	trick /trɪk/	n. 诡计；花招；骗局	
20	observe /əbˈzɜːv/	v. 观察；注视；监控	
21	complicated /ˈkɒmplɪkeɪtɪd/	adj. 复杂的；难懂的	
22	shuttlecock /ˈʃʌtlkɒk/	n. 羽毛球；毽子	
23	badminton /ˈbædmɪntən/	adv. 显然；似乎	
24	artistically /ɑːˈtɪstɪkli/	adv. 有艺术地，在艺术上	
25	vigorous /ˈvɪɡərəs/	adj. 精力充沛的；有力的；元气旺盛的	
26	bow /baʊ/	n. 弓；鞠躬礼 v. 鞠躬；致敬；压弯	
27	arrow /ˈærəʊ/	n. 箭；箭头	
28	aristocracy /ˌærɪˈstɒkrəsi/	n. 贵族；贵族阶级	
29	actively /ˈæktɪvli/	adv. 积极地；活跃地	
30	competition /ˌkɒmpəˈtɪʃn/	n. 竞争；比赛	
31	restrain /rɪˈstreɪn/	vt. 抑制；阻止；束缚	
32	be similar to	与……相似	
32	to point out	指点；指出	
33	to figure out	想出；理解；弄清	

2 Read and learn the key terms of Chinese traditional games，and tick the ones you know in the last column of the word list.

Numbers	Terms	Meanings	Notes
1	checkers	跳棋	
2	board games	棋类游戏	
3	longevity principle	长寿之道	
4	chess set	棋盘	
5	folk game	民间游戏	
6	hand-eye coordination	手眼协调	
7	shuttlecock kicking	踢毽子	
8	bow and arrow	弓箭	

3 Get to know some useful expressions for etiquette of Chinese chess.

According to the basic rules followed by players, there are some etiquette or proverbs in playing Chinese Chess or Go.

(1) A chariot fades many pieces. 一车十子寒。

(2) A horse has an awe-inspiration reputation extending in every direction. 马有八面威风。

(3) A single cannon is hard to capture piece. 孤炮难鸣。

(4) A cross-river solder can act as powerful as a chariot. 卒子过河赛如车。

(5) Use it or lose it. 用之或弃之。

(6) Watching but not telling is a true gentleman. 观棋不语真君子。

(7) Victory and defeat are both common in battle. 胜败乃兵家常事。

(8) Those closely involved cannot see as clearly as those outside. 当事者迷，旁观者清。

Task 3 Consolidated Exercises

1 Answer the following questions briefly according to the passage.

(1) What games do Chinese play for relaxation and mental stimulation?

(2) What is the benefit of shuttlecock kicking?

(3) Why are Chinese traditional games so popular?

(4) Why do some players may hit the board hard at some moves?

(5) What do you think is the reason why Weiqi or Go is more popular in those with higher education?

(6) Why does the author think the game Weiqi or Go is just like life?

(7) What was a very important educational subject in the Western Zhou Dynasty?

(8) Why did Confucius pay much attention to archery?

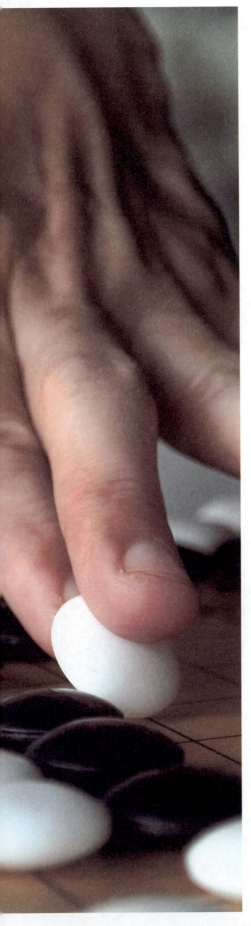

2 Fill in the blanks with the proper words given, and change the word form if necessary.

relaxation	is similar to	keen
principle	maintain	trick
point out	seize	

(1) We are anxious to get a better education so that we can _____ the chance to have a brighter future.

(2) His guiding _____ has been never to stop learning.

(3) A few moments of _____ can work wonders.

(4) He had retained a _____ interest in the progress of the work.

(5) The government struggled to _____ law and order.

(6) The anxiety rate of those men is said to _____ jumping from an aircraft.

(7) He's pulling some sort of _____ on you.

(8) I have to _____ one page is missing in this book.

3 Translate the following sentences into English.

(1) 他很渴望听到不同的声音。

(2) 她不要家人帮忙, 对她来说这是个原则问题。

(3) 我抓住一切机会与外国人交流, 而这加深了我对西方国家的了解。

(4) 人人都需要一些休息和放松的空闲时间。

(5) 正如在上节课中所看到的那样, 这种会议和协商很类似。

(6) 他采用了以攻为守的老招数。

(7) 现今很少有人能够把友谊维持到成年。

(8) 如果您能指出我的不足之处, 我会不胜感激的。

Section III
Let's Practice

Task 1   Let's Talk

1 Listen to the story and write down the proper answers in the blanks.

The Third Shaolin Kung Fu Culture Festival is taking place in London. Kung Fu lovers from all over the world _____(1)_____ the festival at the Shaolin Temple UK, which aims to create an environment for exchanging Chinese culture, religion and _____(2)_____ in Britain. One of the arts greatest fighters Master (大师) Shi Yanzi, who started his Kung Fu journey in the Henan Province in China, set up his own Kung Fu temple in the UK nearly two decades ago. He hopes the festival can bring people together to get a better _____(3)_____ of the sport.

"The main _____(4)_____ for the Third Shaolin Culture Festival is to build up a _____(5)_____ on which all the fans of Shaolin culture in the world, in particular those from Europe can get together at this place and get in _____(6)_____ closely with the Shaolin Culture and then they can have better _____(7)_____ and learn profound knowledge about it."

At the festival students performed various forms and fighting _____(8)_____ representing the many disciplines of this Chinese martial art. The sporting and cultural _____(9)_____ was put together by the venerable (令人尊敬的) Grand Master Shi Yongxin, Abbot (住持) of the Songshan Shaolin Temple. The festival _____(10)_____ until the 14th October.

2 Read and repeat the dialogue.

A: I`ve been learning to play Go recently. It`s an amazing game!

B: Weiqi? It must be a greatest Chinese invention. I`ve tried hard to learn it. Its rules can`t be any simpler, but I`ve never won a game of Go. Well, it`s all Greek to me!

A: At least you have obtained a very basic understanding of the game, haven`t you! I think it is perfect for boosting intelligence, cultivating personality and flexible learning. Every intellectual wishing to gain any insight into Chinese culture should learn to play it.

B: I heard it is included in the four major arts of China, is it?

A: Sure. It was considered desirable that a well-educated ancient Chinese scholar should be well versed in Chinese zither, Weiqi, calligraphy and painting. I think Weiqi is the most fantastic of them. It embodies ancient Chinses wisdom and cultural profoundness.

B: I know it is quite different from Western chess.

A: Yes. Playing chess is a very aggressive experience. All the pieces are supposed to capture their opponents.

B: But isn`t it the same with Weiqi?

A: Certainly not. The object of Weiqi is to surround a larger area than the opponent. In other words, each side is struggling for a greater living space by mapping out a territory on the board. Hence, chess aims to kill, while Weiqi is much concerned with how to survive

Task 2 Team Work

1 Role-play.

Imitate the above dialogue first, then make one with your partner about Chinese Chess or Weiqi and perform it in class.

2 Group discussion.

As we know, it is not easy to keep a shuttlecock in the air. How do you achieve that? Do you have any method to keep it?

Cuju

Cuju (蹴鞠) is an ancient Chinese ball game. It is to kick a ball with feet. "Cu" means kicking with feet, while "ju" refers to a kind of leather ball stuffed with feathers. So combined together, cuju means to play the ball with feet. As a kind of ancient Chinese sport, it is the prototype of the contemporary football.

The form of cuju is similar to that of modern football, but it has many kinds of playing modes and skills, and can be divided into three types: direct-fight, indirect-fight and individual playing.

With a history of about 2,300 years, it started from Linzi (临淄), capital of the Qi State (齐国) during the Spring and Autumn Period (春秋) and Warring States (战国) Period. It achieved great development in the Han Dynasty, and due to its popularity, *25 Articles on Cuju*, a research work on cuju appeared, which is the earliest sports research work in China and also the first professional sports book in the world. The heyday of cuju is in the Tang and Song Dynasties, and the fresco Picture of Playing Cuju on the Horses depicts the scene of the nobles playing cuju on horses. Cuju got huge development in the Song Dynasty. There are some descriptions of the Emperor playing cuju with the court officers in the ancient classic *The Water Margin* (《水浒传》).

Cuju has exerted great influence on modern football. In the Tang Dynasty, Chinese cuju spread to Japan and Korea in the east and Europe in the west, and evolved into modern football in Britain. On June 9, 2004, the Forum on Football Game Originating was held in Linzi District, Zibo City of Shandong Province, and achieved the consensus that ancient cuju originated from Linzi, capital of the Qi State during the Spring and Autumn Period and Warring States Period. On July 15, 2004, on the press conference of the Third China International Football EXPO, FIFA (Federation International Football Association) and AFC (Asian Football Confederation) jointly announced that China is the birthplace of football game, and the world football originated from cuju game in Linzi District, Zibo, Shandong Province of China. In May 20, 2006, as intangible cultural heritage with the approval of the State Council, cuju was included in the first batch of national intangible cultural heritage list. (391 words)

Task 1　New Words and Expressions

Read the following words and expressions, and tick the ones you know in the last column of the word list.

Numbers	Words & Expressions	Meanings	Notes
1	kick /kɪk/	v. 踢；踢腿；踢球得分	
2	leather /ˈleðə(r)/	n. 皮革，皮革制品	
3	feather /ˈfeðə(r)/	n. 羽毛	
4	prototype /ˈprəʊtətaɪp/	n. 原型；样本；规范	
5	contemporary /kənˈtemprəri/	adj. 当代的	
6	popularity /ˌpɒpjuˈlærəti/	n. 普及；流行	

continued

Numbers	Words & Expressions	Meanings	Notes
7	article /ˈɑːtɪkl/	n. 文章；条款	
8	heyday /ˈheɪdeɪ/	n. 全盛期	
9	fresco /ˈfreskəʊ/	n. 壁画	
10	scene /siːn/	n. 场面；情景	
11	noble /ˈnəʊbl/	n. 贵族	
12	description /dɪˈskrɪpʃn/	n. 描述；描写	
13	emperor /ˈempərə(r)/	n. 皇帝；君主	
14	exert /ɪgˈzɜːt/	n. 运用；发挥	
15	forum /ˈfɔːrəm/	n. 讨论；论坛	
16	originate /əˈrɪdʒɪneɪt/	v. 起源；引起	
17	consensus /kənˈsensəs/	n. 一致；合意	
18	press /pres/	n. 新闻界；出版社	
19	conference /ˈkɒnfərəns/	n. 会议，讨论，协商	
20	FIFA /ˈfiːfə/	n. 国际足联	
21	confederation /kənˌfedəˈreɪʃn/	n. 联盟；同盟	
22	jointly /ˈdʒɔɪntli/	adv. 共同地；连带地	
23	announce /əˈnaʊns/	v. 宣布；诉说；预示	
24	batch /bætʃ/	n. 一批	
25	to stuff with	用……填/塞/堵住	

Task 2 Comprehension

Answer the questions after finishing reading the passage.

1. What do "cu" and "ju" mean respectively in traditional culture?

2. How many types does cuju have? What are they?

3. When and where did cuju originate from?

4. When did cuju reach the heyday?

5. What is the name of the fresco picture?

6. How did cuju influence the world?

7. When was football announced its birthplace?

8. When was cuju included in the first batch of national intangible cultural heritage list?

家和萬事興

Insight into Proverbs

"He" 和—Harmony

Referring to peace, "harmony" is regarded as the highest value and moral state in traditional Chinese culture. Happy couples, concordance families, soothing neighbors, a harmonious society, and harmony among nations etc., are the highest realm and goal of the traditional virtues of the Chinese nation.

1. 相互包容，求同存异。

Mutual tolerance, seek agreement while shelving differences.

2. 君子和而不同，小人同而不和。

A wise man is sociable, but not familiar. A fool is familiar but not sociable.

3. 夫唱妇随。

The husband is to sing and the wife is to follow.

（A good Jack makes a good Jill.）

4. 一日夫妻百日恩。

One day together as husband and wife is like a hundred days of grace.

(The marriage relationship is a true blessing.)

5. 天时不如地利，地利不如人和。

The time isn't as important as the terrain, but the terrain isn't as important as unity with the people.

6. 忍一时之怒，可免百日之忧。

If you are patient in one moment of anger, you can escape a hundred days of sorrow (and regret).

7. 远亲不如近邻。

Good neighbors near is better than relatives far away.

8. 君子无所争。

 A gentleman never competes in anything he does.

9. 君子周而不比，小人比而不周。

A wise man is impartial, not neutral. A fool is neutral but not impartial.

10. 家和万事兴。

Harmony in the family brings prosperity.

11. 六亲不和，有孝慈。

It was when the six near once were no longer at peace that there was
talk of dutiful sons.

12. 老吾老，以及人之老；幼吾幼，以及人之幼。

People should extend the respect of the aged in one's family to
that of other families; extend the love of the young ones in
one's family to that of other families.

13. 礼之用，和为贵。

In the practice of art, what is valuable is natural spontaneity.

14. 鹬蚌相争，渔人得利。

When the snipe and the clam fight, it's the fisherman who profits.

(A house divided against itself cannot stand.)

15. 水能载舟，亦能覆舟。

The water that bears the boat is the same that swallows it up.

Section VI
Self-assessment Checklist

1　Now, it's time for you to review your performance after learning this unit. Carry out a self-assessment by checking the following table.

Items	Ratings			
1. Knowledge	A	B	C	D
I know Chinese ancient entertainments and its components.				
I know the basic rules in playing Chinese chess.				
I know the history of traditional Chinese games.				
I know the history of cuju and how it influences the world.				
I grasp the expressions of Chinese ancient games.				
2. Skills	A	B	C	D
I can use logic to grasp the main idea and the details of the passage.				
I can guess the meanings of the new words through word-formation and context.				
I can analyze the ideas and viewpoints of the passage and solve the problems.				
3. Speaking	A	B	C	D
I can tell the types of traditional Chinese games in English.				
I can tell the pieces and its basic rules in playing Chinese Chess.				
I can explain the reasons of some Chinese sayings in playing chess.				
4. Confidence in Chinese Culture	A	B	C	D
I can understand the connotation of Chinese Chess, Weiqi and Tai Chi in traditional Chinese culture.				
I can understand the philosophies in playing Chinese Chess and Tai Chi.				
I feel proud of the connotation of Chinese ancient games.				

A: Basically agree

B: Agree

C: Strongly agree

D: Disagree

2　Fill in the blanks in the mind maps below to check whether you have a good understanding of this unit.

(1) Mind map of Chinese Chess and its pieces.

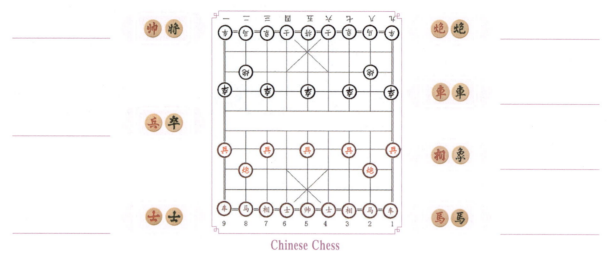

Chinese Chess

(2) Mind map of Chinese ancient entertainments and its types.

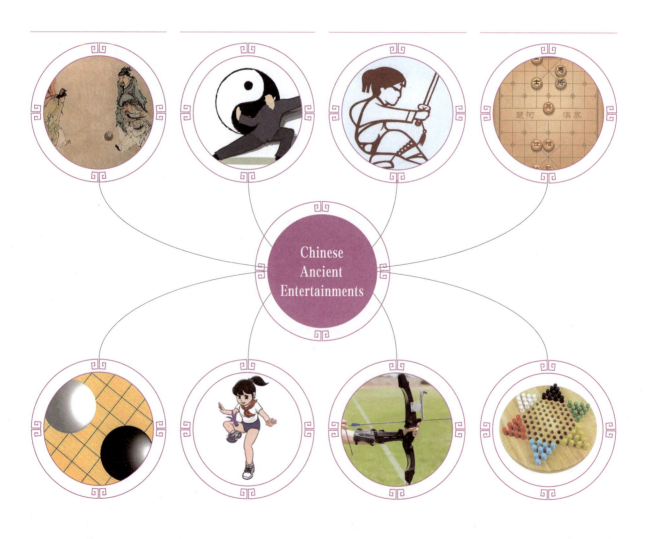